BestMasters

Mit „**BestMasters**" zeichnet Springer die besten Masterarbeiten aus, die an renommierten Hochschulen in Deutschland, Österreich und der Schweiz entstanden sind. Die mit Höchstnote ausgezeichneten Arbeiten wurden durch Gutachter zur Veröffentlichung empfohlen und behandeln aktuelle Themen aus unterschiedlichen Fachgebieten der Naturwissenschaften, Psychologie, Technik und Wirtschaftswissenschaften. Die Reihe wendet sich an Praktiker und Wissenschaftler gleichermaßen und soll insbesondere auch Nachwuchswissenschaftlern Orientierung geben.

Springer awards "**BestMasters**" to the best master's theses which have been completed at renowned Universities in Germany, Austria, and Switzerland. The studies received highest marks and were recommended for publication by supervisors. They address current issues from various fields of research in natural sciences, psychology, technology, and economics. The series addresses practitioners as well as scientists and, in particular, offers guidance for early stage researchers.

Niklas Litzenberger

Introduction of Advanced State Space Grids and Their Application to the Analysis of Physics Teaching

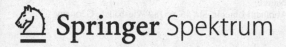
Springer Spektrum

Niklas Litzenberger
Mainz, Germany

ISSN 2625-3577   ISSN 2625-3615 (electronic)
BestMasters
ISBN 978-3-658-42731-3   ISBN 978-3-658-42732-0 (eBook)
https://doi.org/10.1007/978-3-658-42732-0

This Springer Spektrum imprint is published by the registered company Springer Fachmedien
Wiesbaden GmbH, part of Springer Nature.
The registered company address is: Abraham-Lincoln-Str. 46, 65189 Wiesbaden, Germany

Paper in this product is recyclable.

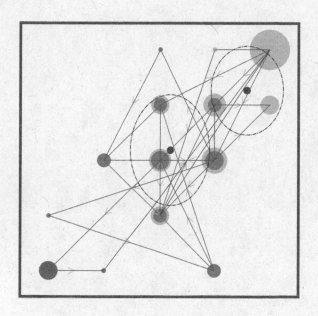

Danksagung

Für die große Unterstützung bei der Erstellung dieser Masterarbeit möchte ich mich bei einigen Menschen bedanken. Mein größter Dank gebührt Andreas Pysik, meinem Erstbetreuer. Er hat sich immer sehr viel Zeit für mich und das Feedback zu meiner Arbeit genommen, wodurch ich vieles verbessern konnte und auf einige neue Ideen kam. Auch mein Zweitbetreuer Dr. Sebastian Wurster, hat mich seitdem ich die Idee der ASSGs hatte immer sehr stark unterstützt und dank ihm konnte ich viele Menschen für die Durchführung meiner Studie gewinnen. Besonders dankbar bin ich Christopher Newton, der sich die große Mühe gemacht hat, den größten Teil meines Textes auf Rechtschreib-, Grammatik und Syntaxfehler zu korrigieren. Auch möchte ich Johannes Lhotzky, Prof. Dr. Klaus Wendt und Prof. Dr. Margarete Imhof für ihr Feedback danken. Nur dank all diesen Menschen ist diese Masterarbeit so weit gekommen, wie sie jetzt ist.

Deutsche Zusammenfassung

Unterricht zeichnet sich durch komplexe Interaktionen aus. Deshalb lassen sich nur schwer globale Aussagen über eine Unterrichtssequenz treffen. Bislang werden jedoch in der indikatorengestützten Unterrichtsbeobachtung häufig Methoden eingesetzt, die auf eine globale Einschätzung des Unterrichts abzielen. Unklar bleibt, wie zuverlässig globale Bewertungen ganzer Unterrichtssequenzen sind, da keine Aussagen über die stattfindenden dynamischen Prozesse gemacht werden können. Diese Schwierigkeit besteht ebenso bei der Analyse von Zusammenhängen zwischen verschiedenen Unterrichtsmerkmalen (z. B. verschiedener Aspekte der Basisdimensionen guten Unterrichts), weil beziehungsweise Korrelationen zwischen Unterrichtsmerkmalen, gemessen über Globalindikatoren, dynamische Prozesse nicht berücksichtigen können.

Ein Lösungsansatz für diese beiden Problematiken sind die von Hollenstein (2013) entwickelten State Space Grids (SSG). Diese ermöglichen den zeitlichen Verlauf der Ausprägung zweier Unterrichtsmerkmale in Beziehung zueinander zu setzen. Dabei werden Bewertungen durch Indikatoren in festen oder auch variable Abständen erfasst und in ein zweidimensionales Koordinatensystem übertragen. Dadurch lassen sich dynamische Prozesse visualisieren. Erste Studien wenden SSGs für die Analyse pädagogischer Interaktionen zwischen Lernenden und Lehrkräften an. Jedoch gibt es hierbei keine geeignete Möglichkeit den Zusammenhang zweier Indikatoren numerisch zu erfassen, um beziehungsweise solche Bewertungen mit Globalbewertungen zu vergleichen.

Ziel der Arbeit ist deshalb, den SSG-Ansatz mit der Erfassung solcher Zusammenhänge zu Advanced State Space Grids (kurz: ASSG) zu erweitern, auf Praktikabilität zu prüfen, mit Globalbewertungen ganzer Unterrichtssequenzen zu vergleichen und ihre Anwendbarkeit auf Physikunterricht zu untersuchen. Hierbei werden mathematisch fundierte, numerische Begriffe und grafische Elemente dem State Space Grid-Plot hinzugefügt. Dazu gehören unter anderem Maße für Streuung der Indikatorbewertungen und Lagemaße die globale Ergebnisse beschreiben.

Die Grundlage dieser mathematischen Maße wird zunächst eingeführt und notwendige Aussagen bewiesen. Darauf aufbauend werden diese neuen Maße der Advanced State Space Grid Methode sukzessiv eingeführt und mit Beispielen erläutert. Mit Hilfe dieser Parameter der neuen Methodik lassen sich bestimmte Typen eines ASSG-Plots unterscheiden. Solche Typen von Plots beschreiben verschiedene Zusammenhänge der aufgetragenen Indikatoren und werden daher Zusammenhangstypen genannt. Diese Unterscheidungen in Zusammenhangstypen werden anhand von numerischen Werten der Parameter festgemacht. Mit Simulationen von Indikatorbewertungen werden diese Werte ermittelt, um Zusammenhangstypen numerisch unterscheiden zu können. Mit Hilfe dieser Werte für die neuen Parameter wird ein Flussdiagramm erstellt, das sukzessiv die Ergebnisse dieser Parameter eines ASSG-Plots durchgeht und darauf basierend eine Kategorisierung in die verschiedenen Zusammenhangstypen liefert.

Um diese neue Methodik auf Praktikabilität und Eignung für Unterrichtssanalyse zu prüfen werden zwei Studien durchgeführt, die Aspekte der Basisdimensionen guten Unterrichts untersuchen. Dabei kann festgestellt werden, dass die ASSG-Methode in weniger Situationen abwendbar ist als herkömmliche Methoden in der indikatorbasierten Unterrichtsforschung. Jedoch ist der Zeitaufwand gleichbleibend und die Menge der Erkenntnismöglichkeiten bei der ASSG Methode ist wesentlich größer als bei herkömmlichen Methoden. Zudem finden sich signifikante Unterschiede bei den Resultaten der Methodiken. Diese Unterschiede lassen sich auf eine geringere Objektivität und Reproduzierbarkeit der herkömmlichen Methoden zurückführen, sowie einen fundamentalen Unterschied in den Untersuchungsmethodiken.

Neben generellen Aspekten der Unterrichtsanalyse gibt es auch Anwendungsmöglich-keiten der ASSG Methode, die besonders für die Analyse von Physikunterricht genutzt werden können. Dazu zählt die Analyse von Interaktionen zwischen Arbeitsblättern und Experimenten, Einflüsse von einzelnen Interaktionen auf den Arbeitsfluss der Lernenden und die Analyse von Unterrichtsstörungen.

Contents

Acronyms

ASSG	Advanced State Space Grid
DC	deviation curve
EV	expected value
ipd	inner point density
MSD	middle standard deviation
SD	standard deviation
SDD	standard deviation difference
SSG	State Space Grid
td	travel distance
tt	travel tendency

Introduction

Lessons are dynamic and consist of complex interactions (Maskus 1976). Such interactions take place not only between students and teachers, but also between students and other students. This poses a methodological challenge for classroom research if this dynamic is to be captured. Research methods often only use global ratings to correlate their indicators. Such global ratings are usually made by handing out rating sheets at the end of each class and comparing them with the results of other classes' ratings (Imhof et al. 2021). Other methods use videos of a class in which a global rating is given for each indicator, substantiated with example situations for each rating (Pretorius et al. 2018). How dynamic interactions affect such global ratings remains unknown, as one cannot analyse dynamic processes by only using global indicators. In particular, the relationships between the basic dimensions of good teaching depend on such dynamic interactions, which are difficult to analyse through global ratings.

A first approach to the analysis of dynamic processes are SSGss (Hollenstein 2013). In this method, all indicators are rated at certain time intervals (e.g. every 60 seconds) so that we can see how the dynamics of the indicators unfold during the observed time. Two of these indicators are plotted in a two-dimensional coordinate system and connected by arrows with increasing time. This allows us to see how the dynamics of the indicators unfold during the time they are examined. Hollenstein then presents various methods and parameters for analysing the dispersion of these points, but they are not based on mathematical principles and do not, for example, provide a concrete and comparable numerical quantity for a global rating of these indicators. The methods for analysing the scatter of the points also depends on the interpretation of the point representation and there is hardly any numerical and stochastic basis for the interpretation.

N. Litzenberger, *Introduction of Advanced State Space Grids and Their Application to the Analysis of Physics Teaching*, BestMasters, https://doi.org/10.1007/978-3-658-42732-0_1

To address these disadvantages of SSGs, a new and expanded version of this research method is being introduced: Advanced State Space Grid (ASSG). In this case, new parameters are constructed that help with the interpretation and are based on stochastic mathematics. Through these parameters, one not only has a basis for further interpretation of whether given indicators are related, but one can also categorise these relations and compare them to other indicator relations. Furthermore, one will then be able to give an indicator an overall result that also takes the dynamic processes into account.

Furthermore, this new ASSG method will be compared with a standard method in the current state of research to see whether the results of these methods differ and what advantages and disadvantages each method has. In particular, it will be investigated whether the global ratings differ from the global results of the ASSG method, as one can then include the dynamic processes in a global result of an indicator, which is not possible with standard methods. Finally, everything that has been learned about this new research method will be summarised in order to understand how ASSGs can be used for the analysis of physics education.

In summary, the research questions of this thesis will be:

a) How can the State Space Grids method be further developed into an Advanced State Space Grid method?

   1) How can SSG plots be statistically described by position and dispersion measures?
   2) How can different types of SSG plots be categorised?

b) How can the ASSG method be applied for the analysis of teaching?

   3) Are there differences in results between the ASSG method and a standard method for indicator-based analysis?
   4) What are advantages and disadvantages of the ASSG method in comparison to a standard method?
   5) How can the ASSG method be used for the analysis of physics teaching?

To answer these research questions, a theoretical background in four different areas is required. The first theoretical background is the current state of research on the basic dimensions of good teaching. This is needed to understand the challenges in this research field. Furthermore, in the first study of a lesson with the ASSG method, a first attempt is made to analyse the interactions between different basic dimensions of good teaching.

In order to examine the differences between a standard method for indicator-based analyses and the ASSG method, a theoretical background of typical research methods is covered in Chapter 3. From these research methods, a suitable method is selected for comparison with the ASSG method.

The third theoretical background needed is a consideration of the SSG method. This is necessary because the ASSG method will build new parameters on the general idea of the SSG method. In order to formulate these new parameters, mathematical foundations are needed. These mathematical foundations are discussed in Chapter 5. In addition to the mathematics needed to formulate the new parameters, other mathematical fundamentals needed to simulate the results for these parameters are discussed.

In Chapter 6, Chapter 7, and Chapter 8 the new parameters will be defined and explained how they can help in categorising the SSG plots. The final objective is to simulate the critical values between these different types of SSG plots in order to create a flow chart that represents a first attempt to categorise these types. This will answer the research question "How can different types of SSG plots be categorised?".

The next two research questions are addressed in Chapter 9 and Chapter 10. In these two chapters, two different studies are compared using the ASSG method and the chosen standard method from Chapter 3 will be compared. In this comparison, the two research questions are answered.

The general applications of the ASSG method for the analysis of physics teaching are also addressed in Chapter 9 and Chapter 10. In these studies, the application of the ASSG method for English and German lessons is tested using indicators that are not specialised for a particular subject. Thus, the general applicability of the ASSG method in analysing physics teaching can also be tested on lessons of different subjects. The application of the ASSG method to specific aspects in the analysis of physics teaching and its specificities are addressed in Chapter 11. This will answer the last aspect of the research question "How can the ASSG method be used for the analysis of physics teaching?".

A conclusion of this thesis will then be given in Chapter 12. In addition, this final chapter also summarises further improvements and other applications of the ASSG method hat are not covered in detail in this thesis.

Basic Dimensions of Good Teaching

2

With the help of the ASSG method, an analysis of dynamic interactions becomes possible. With this method, a gap in the current state of research can be closed, in which only global research methods could be used. One of these gaps is the dynamic interactions between the basic dimensions of good teaching according to Klieme. In Section 2.2 it can be seen that there is hardly any research on the relationships between the different dimensions, as this requires the analysis of dynamic interactions in the classroom, which is not possible or a serious challenge for typical research methods. Therefore, in Chapter 9 a standard research method will be compared with the ASSG method by selecting indicators from the basic dimensions in Chapter 9. In doing so, it can be analysed whether the ASSG method enables an analysis of the relationships between the basic dimensions of good teaching in order to close this gap in the current state of research.

2.1 Basic Dimensions and Sub-Dimensions of Good Teaching

Classroom research is fundamentally divided into three characteristic areas: classroom management, student support, and cognitive activation (Jentsch et al. 2019). This model has the advantage that the effect of teaching and its characteristics can also be explained theoretically (Lipowsky & Bleck 2019). These three characteristics are referred to as the basic dimensions of good teaching (Klieme et al. 2001). Those three aspects have been determined by numerous studies on good teaching, all leading to the same three aspects of good teaching (eg. Lipowsky 2019, Ditton 2008, Hattie 2009, Seidel & Shavelson 2007).

© The Author(s), under exclusive license to Springer Fachmedien Wiesbaden GmbH, part of Springer Nature 2023
N. Litzenberger, *Introduction of Advanced State Space Grids and Their Application to the Analysis of Physics Teaching*, BestMasters,
https://doi.org/10.1007/978-3-658-42732-0_2

Kounin (2006) divides effective classroom management into the following sub-categories: Teacher omnipresence, overlap, smoothness and momentum of instruction, and mobilisation of the learning group. These subcategories are also referred to as sub-dimensions of the corresponding basic dimension. Based on Kounin's classification, this results in guidelines for the teacher's behaviour to prevent disruptions, so that an optimal use of learning time and a structural framework for successful learning processes can be created.(Helmke 2012, Seidel & Shavelson 2007).

The second basic dimension, student support, takes into account Ryan and Deci's (2000) or Rakoczy's (2008) self-determination theory and addresses the extent to which teaching should address students' basic motivational needs of students (Jentsch 2019). In this context, supportive classroom climate is also often referred to, which is avoided by some research groups due to the inconsistent conceptualisation of the climate term (Lipowsky & Bleck 2019). Supportive classroom climate is a multi-faceted construct with multiple conceptualisations and operationalisations (Gruehn 2000). On the one hand the focus is on the type and quality of the relationship between the interaction partners (teacher-pupil relationship or pupil-pupil relationship), but also on the quality of the professional-adaptive support given to the pupils in the learning process (Adelman & Taylor 2005, Gruehn 2000).

According to Baumert et al. (2010), cognitive activation is attributed to problem-oriented learning situations that should lead to a conceptual understanding of the learning object. Lipowsky and Bleck (2019) also emphasise that, in contrast to the other basic dimensions, cognitive activation focuses on a deeper engagement with the learning object. Furthermore, the current state of research shows that cognitive activation exhibits a high degree of variability compared to the other basic dimensions in the lessons under consideration. Accordingly, cognitive activation varies for one and the same teacher depending on the content and goal of the lesson (Klieme & Rakoczy, 2008).

Summed up these three dimensions can be illustrated in pillars of good teaching as in Fig. 2.1:

To further categorize the basic dimensions of good teaching, sub-dimensions are added to each dimension, as in Kounin's definition of classroom management. These sub-dimens-ions differ slightly in different studies (Gabriel 2014, Lipowski et al. 2018, Lipowsky & Bleck 2019), e.g. the sub-dimensions according to Syring (2017), which also includes intellectual challenge for all students. Sub-dimensions of student support could for example be positive affective reactions (praise) and appreciation of classroom contributions.

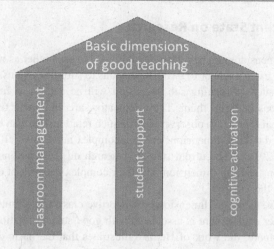

Figure 2.1 Pillars of good teaching (Klieme et al. 2001)

This three-pillar model of teaching quality has already been widely received and empirically confirmed. Nevertheless, a clear tendency to supplement the model can be observed in recent years (Lipowsky & Bleck 2019, Schlesinger et al. 2018, Praetorius et al. 2020). At the center of these efforts is the discussion of whether the dimension of cognitive activation can represent subject- and content-specific aspects in a differentiated enough way. It is often not taken into account which contents are dealt with and linked in the lessons. In addition, the content is often not checked for comprehensibility or subject-specific correctness with regard to cognitive activation. A fourth pillar should deal with the subject-didactic quality of teaching. This should include aspects of subject didactic depth and relevance on the one hand, and aspects of content and structural clarity on the other (Lipowsky & Bleck 2019). Whether this potential new basic dimension (focusing on relevant content and aspects of content and structural clarity) is sufficiently different from the existing basic dimensions or whether additional basic dimensions should be introduced is still unclear according to current state of research. Since the aim is only to investigate whether there are relationships between these dimensions, it is not relevant whether there is an additional basic dimension or not. Therefore, only the first three basic dimensions will be dealt with, as shown in Fig. 2.1 in the following chapters.

2.2   Current State on Research

These basic dimensions can be related or correlated to each other as well as to sub-dimensions (to each other) or to basic dimensions to sub-dimensions. In Chapter 9 indicators for such correlating sub-dimension will be chosen to further analyse possible relations between them. These indicators are chosen to test the ASSG method with real lessons, to observe whether such relationships can be found. It is also emphasised that all dimensions create a complex network of effects (Helmke 2007, Gabriel 2014). The current state on research on the relations between the basis dimensions and sub-dimensions and their complex network of effects will be summarised in this chapter.

For example, the basic dimension of supportive classroom climate has shown that high learner engagement is associated with good student relationships (Mayr 2006). Furthermore, the work of Helmke illustrates that the basic dimensions of effective classroom management and classroom climate strongly influence each other (Gabriel 2014, Helmke 2007). Likewise, individual sub-dimensions that are directly related can balance each other out (Gabriel 2014, Helmke 2007).

Gabriel (2014) highlights another connection between a basic and an unrelated sub-dimension: Adherence to rules is related to the safety of a classroom climate conducive to learning (Pietsch 2010, Gabriel 2014). Likewise, planning ahead and the associated optimisation of learning time is seen as an influencing factor of constructive learning support (Gabriel 2014). Gabriel (2014) thus describes the optimal use of time as a sub-dimension of effective classroom management and therefore as a necessary variable for the basic dimension of learning support (ebd.). Significant for these studies is above all the realisation that there are connections between the individual basic dimensions and sub-dimensions that Mayr discusses: Deficiencies in one area of teacher behaviour can be at least partially compensated for by strengths in other areas (Mayr 2006).

Neuenschwander also addresses the basic dimension of classroom management. Classroom management is not only likely to influence students' teacher images, but also to contribute to an better classroom climate (Neuenschwander 2006). The basic dimensions of a conducive classroom climate and effective classroom management are thus presumably directly linked. Gabriel (2014), on the other hand, says that the results of observational studies to date show low correlations between effective classroom management and a positive classroom climate. Lipowsky and Bleck (2019) and Lipowsky et al. (2019) also do not attribute a favourable relationship to classroom climate with other sub-dimensions and learning success.

This comparison illustrates that the subjectivity of the raters significantly influences the data, as the focus and perspective vary individually (Bortz and Döring 2016). The structured interaction of individual basic and sub-dimensions ensure short- and long-term learning successes, which essentially constitute good teaching (Gabriel 2014, Klieme et al. 2001). The ASSG method attempts to reduce the influences on the data by the subjectivity of the researcher. This will further be compared with a single global rating in Section 10.3.

Research Methods

3

In Chapter 9 the ASSG method will be compared with a standard research method in order to answer the research questions "Are there differences in results between the ASSG method and an standard method for indicator based analysis?" and "What are advantages and disadvantages of the ASSG method in comparison to an standard method?". Since different research methods are used, this chapter aims to provide an overview of the types of research methods that are used. With this overview, the ASSG method can be placed in the context of the already existing methods.

3.1 Evaluation Sheets

Evaluation sheets are a commonly used method for analysing lessons. They consist of a series of questions that can be answered by selecting options on a scale. A commonly used scale for such questions are Likert scales (Imhof et al. 2021). They provide a "a range of responses to a given question or statement" (Cohen et al. 2000, p. 313). "Typically, there are five categories of response, from (for example) 1=strongly disagree to 5 =strongly agree, although there are arguments in favour of scales with seven, or with an even number of response categories" (Jamieson 2004, p. 2). Through these questions a researcher can measure the attitude of each individual to the subject of the question.Since the questions can easily be misinterpreted, each aspect is divided into subcategories of different questions, this is also the case, for example, in Likert scales (Imhof et al. 2021).

Supplementary Information The online version contains supplementary material available at https://doi.org/10.1007/978-3-658-42732-0_3.

N. Litzenberger, *Introduction of Advanced State Space Grids and Their Application to the Analysis of Physics Teaching*, BestMasters,
https://doi.org/10.1007/978-3-658-42732-0_3

These scales have a level of measurement because "the response categories have a rank order, but the intervals between values cannot be presumed equal" (Jamieson 2004, p. 2). Even if this is not the case, "researchers frequently assume that they are" (Blaikie 2003, p. 24) equal. Therefore "treating ordinal scales as interval scales has long been controversial" (Jamieson 2004), because the use of ordinal scales for parametric analyses where the distances between values are not equal is one of the "Seven Deadly Sins of Statistical Analysis" (Kuzon et al. 1996, p. 265). This effect is further intensified by the polarisation and tendencies of the individual subjects, as they have certain reaction tendencies that differ from person to person (Imhof et al. 2021).

Evaluation forms provide good insights into the subjective opinions of individuals, however they cannot be used for statistical analysis. As a result they can therefore not even be used to form averages, which they are at the moment.

3.2   Indicator-based Measurements

Indicators provide a different approach to collecting data for an investigation. In particular, the goals of learning analytics "(e.g. monitoring, analysis, prediction, intervention, tutoring, mentoring, assessment, feibidack, adaptation, personalization, recommendation, awareness, reflection) need a tailored set of indicators and metrics to serve different stakeholders with very diverse needs" (Chatti et al. 2004, p. 14). Indicators "can be described as specific calculators with corresponding visualizations, tied to a specific question" (Dyckhoff et al. 2012, p. 60). For example "the fuel needle of a car is an indicator" (Glahn 2009, p. 56).

In a lesson context "The teacher moves around the classroom" is an indicator which will be used in Chapter 9 to compare the ASSG approach to a standard method in the analysis of lessons.

These indicators can then be further described through rankings (ebd). In the case of the indicator from Chapter 9 this would be like this:

- A 5 is awarded if the teacher completes four or more position changes.
- A 4 is awarded if the teacher completes three position changes.
- A 3 is awarded if the teacher completes two position changes.
- A 2 is awarded if the teacher completes one position change.
- A 1 is awarded if the teacher completes no position change.

This indicator can lead to different results if the time at which it is measured varies, which is why it is necessary to set a specific time interval at which it is measured. In this example, the measurement was set to every two minutes.

A measurement such as in the example above could also be done by only counting all position changes over the total time of the whole lesson. A distinction is thus made between low-inferential indicators and high-inferential indicators. With low-inferential indicators, the subjective interpretation of the observer can have a greater influence, the degree of inference by the observer is kept to a minimum and the measurement time for each assessment is small (Lotz et al. 2013). The indicator above would be an example of such an indicator with low interference, as the measurement can be counted and the influence of the rater is low. In comparison, high-inference indicators have low rater influence and can therefore be measured at longer time intervals, e.g. only at the end of a whole lesson (ibid.).

Especially low inferential indicators are highly dependent on the rater, which can be somewhat compensated for by rater training, but still represents a major hurdle for indicator-based analyses (Döring and Bortz 2016). Rater training is also carried out in the indicator-based analyses in Chapter 9 and 10 and can furthermore be found in chapter E of the electronic supplementary material.

This method described by Lotz et al. is intended to serve as a comparative method between the ASSG method and an already established method in classroom research. The reason for this is that the same indicators as in the ASSG method can be chosen and thus a much better comparability is given. Since this method does not have a name it will be called the "standard method" in this thesis.

3.3   Thin-slicing

Since only small periods of time are evaluated for low-inferential indicators, the idea was born to evaluate indicators only in small sections in order to obtain statements about the global, i.e. total, teaching time. This idea to "slice" a total teaching time into thin pieces is called "thin-slicing" (Ambady et al. 2000).

"A thin slice is an excerpt of behavior that is shorter than the total duration of the behavior the researcher has at hand, whether it consists of video, audio, or transcribed text" (Wang et al. 2021, p. 54).

The length of each slice varies from researcher to researcher. Whilst some use $50\ ms$ for one slice (Rule and Ambady 2008) others take $10\ min$ for a single slice (Hirschmann et al. 2018). The number of slices taken into consideration also varies from a single slice to a large number of slices (eg. Scherer 1972; Ambady and Gray 2002; Ambady and Rosenthal 1993; Fowler et al. 2009; Houser et al. 2007; Kraus and Keltner 2009; Oltmanns et al. 2004), as does the percentage of time observed compared to the total time of the observation base (eg. Levine and Feldman 1997; Goh et al. 2019).

A comparison between a single slice and combined slices totalling 5 minutes shows that rating a single slice leads to inferior results compared to a 5-minute slice. The position of the slice does not change this either, and there is no optimal position for a single slice (Wang et al. 2021). However, a comparison of cumulative slices with total time shows that a sufficient number of slices can predict the outcome of a global rating, even if the time covered by the slices is shorter than the total time (ibid.).

3.4   Field Experiment

Another method is a field experiment. Here, research questions are formulated as testable predictions about the theoretically expected outcome of the data collection (Imhof et al. 2021).The data are then collected in an experiment with different groups in their usual environment. A laboratory experiment is similar to a field experiment, where data is collected in an artificial environment. Laboratory and field experiments differ in two topics. "The first is whether results from the experiment can be generalized, and the second is the level of control that the experimenter has. These themes represent a trade-off, since generalizability can sometimes be higher in the field, but this comes with some loss of control" (Samek 2009, p. 104). Especially for the analysis of lessons, field experiments in a real classroom offer greater generalisability if the content-related question focuses on teaching-learning forms in schools (Imhof et al. 2021).

The method of data collection itself can be freely chosen (ibid.). Both evaluation forms and indicator-based measurements would be conceivable as implementation options. Since both example analysis for the ASSG-method (see Chapter 9 and 10) use videos of lessons in a classroom for data collection, both measurements can be considered as a field experiments. However, in this thesis, as already described, only indicator-based measurements are relevant.

3.5   Correlation Studies

In the previous chapter, a traditional experimental research method and its signifi-cance for this thesis were described. The central advantage of such a method is that it enables causal statements about approximate causality (ibid.). However, for many questions it is not possible to carry out an experimental manipulation. For example a "correlation between self-efficacy and job satisfaction of teachers" (Yildirim, I. 2015, p. 478) or the "correlation between the practical aspect of the course and the e-learning progress" (Bylieva et al. 2009, p. 1) are aspects that are difficult to carry out in a field experiment. This remains unchanged even with a large number of participants. This is down to the fact that one would need a large number of experiment groups that would furthermore have to be observed over a large time period in order to obtain a sufficient data set.

On the other hand, if one wants to know the extent of the impact of a supportive parenting style on a child's academic performance, one cannot use an experimental design either. It would not be ethical to divide parents into an experimental group in which they support their child and a control group in which they do not (Imhof et al. 2021). In such a situation, one is dependent on observing which parenting style parents use on their own and to what extent the academic performance of their children is related to this (ebd.). In this way, indirect data can be obtained to further analyse the relationships of the research question. Data collection can be done through evaluation forms or indicator-based measurements. One could also use already existing data from other researchers (Praetorius 2018). The correlation itself is then analysed using statistical methods such as the a χ^2-test which is explained in more detail in Section 7.1.1.

The ASSG method can be used in correlation studies as well as in field experiments. The application of the ASSG method in field experiments is dealt with in Chapters 9 and 10, whereas the usage in correlation studies is only explained in Chapter 12.

State Space Grids

4

As shown in Chapter 3, there are various methods to determine the result of an indicator or the answer to a question for the whole duration of a lesson. However, a method for the analysis of dynamic processes is not included. The first approach to such a method is given by Hollenstein with State Space Grids, which is examined in more detail in this chapter. In Chapter 6 this method will be enhanced with the mathematical statistics of Chapter 5.

4.1 Dynamic Systems

The interaction between students and the teachers have already been summarised and analysed in many studies, as can be seen in Section 2.2. However, a closer look at moment-to-moment interactions is yet to be done. Pennings and Hollenstein, on the other hand, take a different approach. "The present study combined insights from interpersonal theory and dynamic systems approaches to study indices of interpersonal content and structure in teacher-student interactions" (Pennings & Hollenstein 2019, p. 2). This type of approach involving dynamic systems can be realised with the help of state-space grids.

Dynamic systems are divided into micro, meso, and macro systems. Microsystems are very small systems in which the components are individual objects. It is used to describe the interaction of these particles. An example from school would be the interactions between individual students. Macrosystems, on the other hand, describe macroscopic components that depend on multiple factors. In school, this would be the entire organisation of the school. There are interactions between individual class-teacher systems, entire student levels, and the teaching staff. A macrosystem describes the interactions between these systems. "In between these two is an

© The Author(s), under exclusive license to Springer Fachmedien Wiesbaden GmbH, part of Springer Nature 2023
N. Litzenberger, *Introduction of Advanced State Space Grids and Their Application to the Analysis of Physics Teaching*, BestMasters,
https://doi.org/10.1007/978-3-658-42732-0_4

intermediate one, here labeled as ‚meso'." (Hollenstein 2013, p. 4). The following illustration highlights the relationships between micro, meso, and macro systems.

Individual points in a micro system are the students in the school example. The mesosystem then describes the interaction in class between the teacher and the students and, if necessary, further interaction partners, such as a visiting parent. However, such further interaction partners are not taken into account in this work. However, they could also be described with the help of state space grids. Systems that describe interactions between mesosystems are called macrosystems, as in the case of interactions between classes and the teaching staff. The following chapter will answer what such a mesosystem looks like.

4.2   Parameters of State Space Grids

The method of state space grids is used in particular in psychology and student-teacher interactions to describe and analyse the interaction between parents and children.

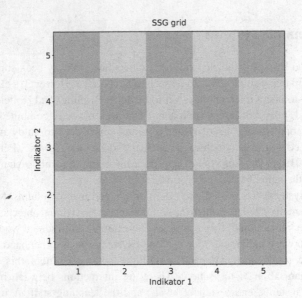

Figure 4.1 Grid of a 5x5 State Space Grid

(Hollenstein 2013, p. 16). State Space Grids (SSGs for short) are two-dimensional systems. They have two axes that can assume discrete values. Thus, a two-dimensional grid is spanned. How this grid can be described is explained in detail in Section 5.1. An example of such a grid can be seen in Fig. 4.1. As can be seen there, variables are assigned to the two axes and their possible values are plotted. The size of a grid thus depends on the values a variable can assume. Therefore, the grid with two variables, each of which can assume four values, is called 5x5 State Space Grid. Similarly, the State Space Grid from Fig. 4.2 is an 8x8 State Space Grid.

On this grid of the state space grids, Hollenstein defines different parameters such as attractors or repellors. The most frequently used parameters are described in the following chapter. A complete list of the existing parameters can be found in "State Space Grids" by Hollenstein 2013 beginning on page 67.

4.2.1   Attractors and Repellors

The interactions between teachers and students have already been described and analysed in some studies through the use of state-space grids (cf. Pennings 2014, Mainhard et al. 2011, Scherzinger et al. 2020 and Pennings & Hollenstein 2019). One possible example of such state space grid of teacher-student interaction can be seen in Fig. 4.2. The relationship between two characteristics of teachers has also been analysed. The "dimensions of dominance vs. submission and hostility vs. affection [...] are utilized to describe how teachers and students relate to each other in class" (Mainhard et al. 2011, p. 1028).

If one looks at the ratings in the SSG from Fig. 4.2, the first circle for a rating in the coordinate $(1\ Assured,\ 7\ Confrontational)^T$ can be seen. In this case let $1\ Assured$ be the value of the variable $Teacher$ and $7\ Confrontational$ the value of the variable $Class$. This notation is discussed in more detail in Section 5.1. This circle serves as the starting value. The temporal progression of the ratings is represented by their connection through arrows. The last value is therefore $(1\ Helpful,\ 6\ Dissatisfied)^T$. The radius of the circle describes how long this state lasts. Large dots represent long-lasting states and small dots represent short-lasting states.

With the help of these dots and arrows, one can follow the dynamic progression of the ratings over the entire duration of the lesson. In this case, the ratings start with a high value for both indicators. Then, the value for the indicator "Class" decreases form $7\ Confrontational$ to $6\ Dissatisfied$ while the other indicator remains the same. Next, the value of the indicator "Class" decreases by three ratings. At the same time, the indicator "Teacher" increases by 6 ratings. After that, the value for the

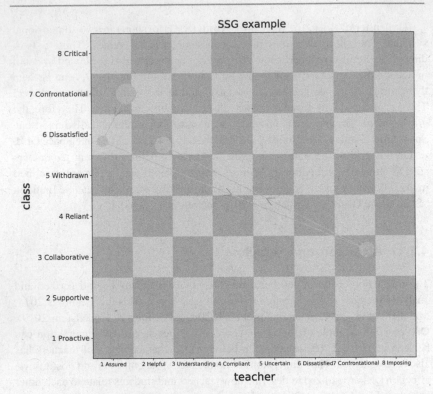

Figure 4.2 Example of State space grid for interaction between students and teacher with indicators of Pennings and Mainhard (2016)

"Teacher" indicator goes back to *2 Helpful* while the indicator "Class" goes back to *6 Dissatisfied*. In this section of the grid, the values remain until the end.

It can be observed that some SSGs tend to have many values in the same coordinate over time. This is the case in the SSG plot in Fig. 4.2 for the lower ratings for the indicator "Teacher" and the higher ratings for the indicator "Class". These "recurrent states are called attractors, because they ‚pul' the trajectory of the system toward those states" (Hollenstein 2013, p.6).

In contrast, the values outside of the area of an attractor are called repellors (ibid.). A repellor can be found in the location $(8\ Confrontational,\ 2\ Collaborative)^T$. There the SSG ratings are outside the attractor and also return to the attractor after some time.

Areas around an attractor are called stabilities. In Section 6.5, a numerical quantity describing this area is defined for this purpose. If an SSG has several attractors, the regions around the attractors are called multistabilities.

4.2.2   Variability

Attractors and repellors can be used to describe the general location of the ratings in the SSG plot. However, the SSG method should also describe the dynamics of the processes in a student-teacher system. This can be done through the variability of the system, which describes how variable the ratings of the indicators are. If the indicators jump frequently form one location to another, the variability is higher than in a system where the ratings stay at one value through the whole time of the lesson.

The difference between high and low variability becomes evident in the following figures. Fig. 4.3 shows an SSG plot with lower variability than in the SSG of Fig. 4.4.

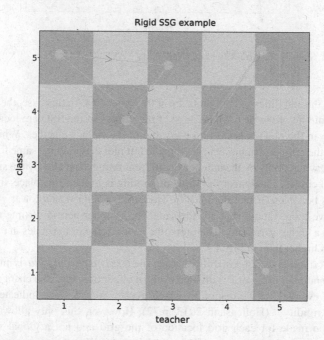

Figure 4.3 State Space Grid with lower variability

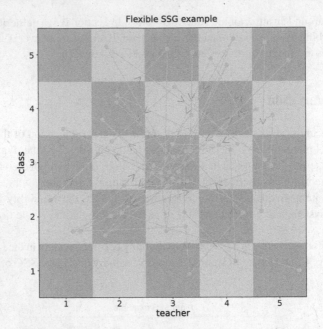

Figure 4.4 State Space Grid with higher variability

However, the stabilities in both plots are at the same area. In this case, the stability is in the middle of the plot, where most of the ratings of the SSG are located. The difference in the plots is how often the value of the ratings changes. While in the right plot the ratings change very often, in the left plot they stay the same for longer. This is shown not only by the number of different ratings, but also by the size of the individual circles. The longer the value of a rating is in the same place, the bigger the radius becomes (Hollenstein 2011). Therefore, a SSG with a lower variability has to have more ratings that are the same for a longer period of time than in a SSG with a higher variability. Therefore, the general size of the cicles in the grid in Fig. 4.3 is higher than in Fig. 4.4.

To describe a low or high stability, the parameter *visits/Event Ratio* is introduced, where the number of "visits" is divided by the number of interaction changes. "With repeating events […] the degree to which events exceed visits indicates ‚stuckness' or ‚rigidity'" (Hollenstein 2013, p.72). However, this only allows a statement to be made for each grid location of the grid and not a global statement about the entire grid. This categorisation is therefore replaced by new parameters in

Chapter 6. With these parameters it is also possible to determine the degree of dispersion and density around an *attractor*, which is not possible with the *visits/event ratio*. This also makes it easier to distinguish between high and low variability.

Hollenstein also introduces two further parameters to describe the type of variability ("State Space Grids" by Hollenstein 2011 starting on page 67). However, these two parameters will be not integrated in the ASSG method, as other more appropriate parameters for describing variability are introduced in Chapter 6.

4.3   Applications for Teaching Anaylsis

4.3.1   Data Collection

One way of collecting data would be classroom videos. Here, indicators are defined and evaluated over the total time of a lesson. Indicators are defined characteristics whose fulfilment is evaluated with the help of rankings (see Secct. 3.2). The indicator has to fulfil certain requirements in order to be used for a SSG. The indicator must be assembled at each time interval of the video. For example, student activity indicators can always be rated as long as there are students in the video sequence. If one then evaluates the students by counting those who fulfil the observable characteristic, one obtains a precise number that clearly leads to an evaluation of the indicator. However, this method depends on the subjective impressions of the researcher, as it is not always clear whether the students fulfil the trait being studied. Therefore, objectivity depends heavily on the choice of indicator and the uniqueness of the trait being studied. For example, it is difficult to assess whether certain students are following the lesson, as they may do so even if they are not looking at the teacher or looking at their worksheet. However, if one chooses an indicator that determines how many students are looking at the teacher or their worksheets, one can assess this characteristic in greater clarity. According to the results of Mayr and Neunschwader from Section 2.2, it should be possible to show such a correlation of two indicators from the basic dimensions in a lesson video. If the indicators are used as variables for a SSG, two indicators can be plotted against each other. To do this, constant time intervals must be chosen in which the indicators are evaluated. This results in discrete values from one to four for each time interval, which can then be plotted against each other.

4.3.2 Possibilities and Limitations of the SSG model

If the indicators are interconnected, as assumed in Section 2.2 they should now be in a state of stability. According to Hollenstein (2013) the indicators correlate more if the ratings in the plot are more concentrated at one particular rating. *Repellors* that bring a system out of stability serve as events that are slight disturbances. Since the time interval leading to a *repellor* is known in an SSG, disruption of stability can be timed there. Events that then cause the system to transition back to stability can then be interpreted as countermeasures for the disruption. Within the SSG, it can then be observed how long stability lasts after the first *repellor*. How long the stability is held thereafter then provides an indication of how sustained the countermeasure was. To improve the lesson in terms of indicators, one can look at what leads to a dispersion of stability leading to a higher value of the indicator. If one could reproduce this, it would be possible to shift a stability in a desired direction to improve instruction. However, the SSG method does not provide a parameter to measure such an interaction. It can only be described by the position of the ratings and is therefore subject to the interpretation of the researcher. To overcome this drawback, a way to measure the magnitude of the countermeasure is introduced in Section 7.4.

Mathematical Fundamentals

5

As became evident in the last chapter the SSG method has its shortcomings. In Chapter 6 new parameters are introduced to address these shortcomings. These parameters are based on mathematical statistics and will be simulated in Chapter 8. Therefore, this chapter serves as an introduction to these mathematical fundamentals which are used in the new ASSG method and the simulation of its parameters.

5.1   Vectors in Cartesian Coordinates

First the grid of an state space grid will be described mathematically. As can be seen in Fig. 5.1 a state space grid consists of a nxm grid in a two dimensional system.

Here n and m are elements of $\mathbb{N}$. It is not necessary that $n = m$. The values on each axis represent the value of the ranking for the indicator that is applied to it. Therefore, the maximum size of the axis is equal to the maximum value of the ranking. For example, an indicator with a possible ranking from one to eight requires a grid size of eight to be able to apply the values on the axis. Should this indicator be combined with a different indicator with only four possible rankings a 8x4 grid is needed (Hollenstein 2013). Since all analysed indicators in this thesis have five possible rankings, each grid will be 5x5 in size.

To further calculate the parameters, each value of a particular time is interpreted as a vector. This can be defined as follows:

© The Author(s), under exclusive license to Springer Fachmedien Wiesbaden GmbH,   25
part of Springer Nature 2023
N. Litzenberger, *Introduction of Advanced State Space Grids and Their Application to the Analysis of Physics Teaching*, BestMasters,
https://doi.org/10.1007/978-3-658-42732-0_5

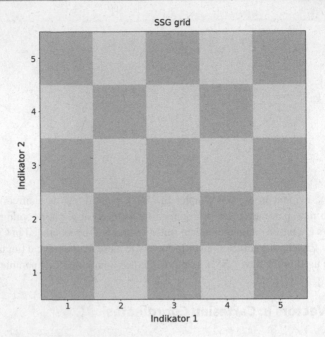

Figure 5.1 Grid of a 4x4 State Space Grid

Definition 5.1.1: Vector of two given indicators
Let $x_i \in \mathbb{N}$ be the value of the ranking of the indicator on the x-axis and let $y_i \in \mathbb{N}$ be the value of the indicator on the y-axis at the given time t_i. As a result the vector v_i of the two indicators is equal to

$$v_i := (x_i, y_i)^T = \begin{pmatrix} x_i \\ y_i \end{pmatrix} = \begin{pmatrix} \text{value of the indicator on the } x\text{-axis} \\ \text{value of the indicator on the } y\text{-axis} \end{pmatrix} \quad (5.1)$$

(Schirotzek 2005).

If, for example, the indicator A on the x-axis and indicator B on the y-axis have a ranking of one and five in the third time section their interpretation as a vector would be $(x_3, y_3)^T = (1, 5)^T$. Such coordinates x_i and y_i described by the lengths on each axis are called Cartesian coordinates (Beetz 2015)

Moreover, the grid in which these vectors lie could be interpreted as a $\mathbb{N}^2$ vector subspace of a $\mathbb{R}_+ x \mathbb{R}_+$ vector space (van Dongen 2015). While the vectors of the indicators can only take on integer positive numbers, calculated parameters can only take on real positive vectors. These parameters can then be represented in a $\mathbb{R}_+ x\ \mathbb{R}_+$ vector space. Therefore, every calculation with vectors takes place in the $\mathbb{R}_+ x\ \mathbb{R}_+$ vector space. Other scalar parameters without a two dimensional calculation can also take negative real numbers, therefore scalar parameters are elements of $\mathbb{R}$.

5.2   Parametric Curves in Polar Coordinates

Instead of defining a vector through values on the x- and y-axis, it is also possible to define a vector through a radius and an angle. These coordinates are called Polar coordinates (van Dongen 2015). This makes it easier to describe a parametric curve like a circle. Since one of the new parameters of the ASSG method is a ellipse, it is easier to describe this curve through the use of Polar coordinates.

The relationship between the polar and Cartesian coordinates can be illustrated as in Fig. 5.2. In this right triangle the following relations apply

$$\sin(\varphi) = \frac{y}{r} \text{ and} \tag{5.2}$$

$$\cos(\varphi) = \frac{x}{r}. \tag{5.3}$$

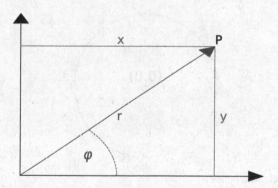

Figure 5.2 Relation between polar and Cartesian coordinates (Papula 2017)

Through the use of equations 5.2 and 5.3 polar coordinates can be defined by Cartesian coordinates as follows:

Definition 5.2.1: Polar coordinates

Let $a \in \mathbb{R}x\mathbb{R}$ be a vector, $r \in \mathbb{R}_+$, and $\varphi \in [0, 2\pi)$, then a can be described through the polar coordinates r and φ with:

$$a = \begin{pmatrix} a_x \\ a_y \end{pmatrix} = \begin{pmatrix} rcos(\varphi) \\ rsin(\varphi) \end{pmatrix} \tag{5.4}$$

(van Dongen 2015).

Using these polar coordinates, a circle can be described by adding points with a fixed radius r and a changing angle $\varphi \in [0, 2\pi)$ (Bär 2010). This leads to a parametric curve

Example 5.2.1: Circle as a parametric curve

Let $r \in \mathbb{R}_+$ and c be a mapping with

$$c : \mathbb{R} \rightarrow [-1, 1] \times [-1, 1]$$

$$c(\varphi) = \begin{pmatrix} r\cos(\varphi) \\ r\sin(\varphi) \end{pmatrix}.$$

Then c describes a circle with radius r and center in $(0, 0)^T$ as in Fig. 5.3 (ibid.).

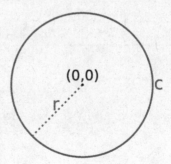

Figure 5.3 Circle with radius r

This example can be further generalised by adding a vector $(m_x, m_y)^T$ to the curve $c(\varphi)$ that shifts the center of the circle to the point of the added vector. An example of this can be seen in Fig. 5.4, where the first circle is moved to the second circle by $(1, 1)^T$.

A second generalisation can be made by varying the radius r of the x- and y-axis independently. This allows the circle to be stretched in the x and y directions. This leads to a parametric description of an ellipse.

Definition 5.2.1: Parametric description of a ellipse
Let $(m_x, m_y)^T \in \mathbb{R}^2$, $(r_x, r_y)^T \in \mathbb{R}_+^2$. Then $E : \mathbb{R} \to \mathbb{R}^2$,

$$E(\varphi) = \begin{pmatrix} m_x + r_x cos(\varphi) \\ m_y + r_y sin(\varphi) \end{pmatrix} \tag{5.5}$$

is a ellipse with center $(m_x, m_y)^T$ (van de Craats & Bosch (2010)).

Example 5.2.2: Stretching of an ellipse
An example for different parameters in an ellipse can be seen in Fig. 5.4. For $r_x = r_y$, the ellipse is a circle because it is not stretched by the difference between r_x and r_y.

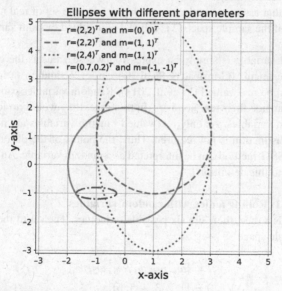

Figure 5.4 Examples for different values for r and m

Values higher than 1 for r_x or r_y increase the area of the ellipse while lower values decrease it. When $r_x > r_y$ the ellipse is stretched in the x-direction (see green ellipse), at $r_x > r_y$ the ellipse is stretched in the y-direction (see red ellipse).

5.3   Distributions

To find critical values for the parameters of the new ASSG-method, different types
to indicator ratings will be simulated. Three different probability distributions are
used for the simulation, which are presented in this chapter.

5.3.1   Discrete Probability Distributions and Random Variables

First of all, a probability distribution has to be defined. A probability distribution
is a function that assigns a probability to every possible individual event. "The set
of possible outcomes is called the sample space" (Grinstead & Snell 2006, p. 18).
Such individual events are called random variables. These random variables can be
defined as follows:

Definition 5.3.1: Random variable
A function X that assigns exactly one number x from the set of real numbers $\mathbb{R}$ to
each result ω of the sample space Ω is called a real-valued random variable (Henze
1997).

"A random variable is simply an expression whose value is the outcome of a
particular experiment" (Grinstead & Snell 2006, p. 1). A random variable does not
always have to be real valued (Li et al. 2013). Random variables can also take on
other values outside their real numbers. In this case, they would no longer be real
valued random variables. As only real valued random variables are needed in this
chapter, this distinction is not required. Thus, the points of the different indicator
ratings in the SSG-method can be interpreted as a random variable. An example can
help understand this definition.

Example 5.3.1: Rolling a dice with random variables
When one rolls a dice, there are six possible outcomes. The set of these outcomes
form the sample space

$$\Omega = \{\omega_1, \omega_2, \omega_3, \omega_4, \omega_5, \omega_6\}$$
$$= \{1, 2, 3, 4, 5, 6\}.$$

The values of an experiment in which a dice is thrown fours times would therefore be

$$X_1, X_2, X_3, X_4.$$

These variables X_i are random variables of the event in which a dice is rolled once. Probabilities are defined by the probability density function $f(X)$ and the probability measure $\mathbb{P}$ of the given random variable X. For example the probability of rolling the number 3 on the first roll of the dice would be

$$f(\omega_1) = \mathbb{P}(X_1 = \omega_3) = \frac{1}{6}.$$

As seen in the example above, the probability density function assigns a probability to each possible outcome.

Definition 5.3.2: Probability density function
Let X be a random variable, then

$$\rho : \Omega \rightarrow [0, 1]$$
$$\rho(x) - \mathbb{P}(X = x) \tag{5.6}$$

is the probability density function of the random variable X (Eichler & Vogel 2011).

The probability density function applies only one probability to a single outcome of the sample space. For a set of outcomes, the probability distribution is needed.

Definition 5.3.3: Probability distribution
Let X be a random variable with the probability density function $f : \Omega \rightarrow [0, 1]$. Then

$$F : \mathcal{P}(\Omega) \rightarrow [0, 1]$$
$$F(x) = \mathbb{P}(X \leq x) \tag{5.7}$$

is the probability distribution of X (ibid.). $\mathcal{P}(\Omega)$ is the power set, which is the system of all subsets of Ω (Henze 1997).

In order to quickly summarise all the necessary objects to describe a probability experiment, the probably space is particularly convenient. By using the probability space, one can abbreviate all previous definitions as follows:

Definition 5.3.4: Probability space and measurement space
The triple of the sample space Ω, the power set of the sample space $\mathcal{P}(\Omega)$, and the probability measure $\mathbb{P}$ is called the probability space $(\Omega, \mathcal{P}(\Omega), \mathbb{P})$ and $(\Omega, \mathcal{P}(\Omega))$ the measurement space.

Utilising the probably measurement $\mathbb{P}$, a distinction can now be made between independent and random dependent. This applies to random variables for which the result of another random variable depends or does not depend on the result. This distinction will be necessary for future definitions and theorems.

Definition 5.3.5: Independent random variables
Let $(\Omega, \mathcal{P}(\Omega), \mathbb{P})$ be a probability space and X, Y be random variables. X and Y are independent random variables, if for any $B, C \subseteq \mathcal{P}(\Omega)$ the following holds true:

$$"P(\{X \in B\} \cap \{Y \in C\}) = P(\{X \in B\}) P(\{Y \in C\})" \tag{5.8}$$

(Behrends 2012, p. 135).

Through the use of this new concept, further calculation can be carried out with the previous example.

Example 5.3.2: Rolling a dice with probability distributions
Continuing the example from earlier, one can now calculate the probability of rolling a number below 5 with a dice. This result is called A. As a set A can be described using $A = \{\{1\}, \{2\}, \{3\}, \{4\}\}$. The subsets of A can be interpreted as independent random variables and can therefore be calculated independently of each other. Then the probability of A is

$$\begin{aligned}
F(A) &= \mathbb{P}(A) \\
&= \mathbb{P}(x \le 4) \\
&= \mathbb{P}(x = 1) + \mathbb{P}(x = 2) + \mathbb{P}(x = 3) + \mathbb{P}(x = 4) \\
&= \frac{1}{6} + \frac{1}{6} + \frac{1}{6} + \frac{1}{6} \\
&= \frac{2}{3}.
\end{aligned}$$

As the example suggests, the probability of an event depends on the associated probability measure of the random variable. Since each probability measure is defined through the distribution applied to the random variable, it is necessary to apply different distributions to simulate different indicator rankings.

5.3.2 Discrete Uniform Distribution

The simplest distribution is the uniform distribution. With this distribution, the same probability applies to each result of the sample space. If the sample space is not a continuous set but a discrete set, the discrete uniform distribution can be used.

Definition 5.3.6: Discrete uniform distribution
Let $(\Omega, \mathcal{P}(\Omega))$ be a measurement space and A be an outcome. Then

$$\text{``}\mathbb{P}(A) = [\ldots] = \frac{|A|}{|\Omega|}\text{''} \ (ibid, p.\ 45). \tag{5.9}$$

is the probability measure of the discrete uniform distribution.

$|\Omega|$ is the number of elements in the set Ω, likewise $|A|$ is the number of elements in set A.

With the discrete uniform distribution the example form above can be calculated too.

Example 5.3.3: Rolling a dice with the discrete uniform distribution
Let $(\Omega, \mathcal{P}(\Omega), \mathbb{P})$ be the probability space of the discrete uniform distribution with one dice. The probability of set $A = \{\{1\}, \{2\}, \{3\}, \{4\}\}$ can be calculated as follows

$$\mathbb{P}(A) = \frac{|A|}{|\Omega|} = \frac{\left|\{\{1\}, \{2\}, \{3\}, \{4\}\}\right|}{\left|\{\{1\}, \{2\}, \{3\}, \{4\}, \{5\}, \{6\}\}\right|} = \frac{4}{6} = \frac{2}{3}$$

which leads to the same result as before. Since every probability of every single outcome in the sample set Ω is the same, every probability of the density function is also the same. This results in the probability density function of a discrete uniform distribution which is shown in Fig. 5.5. Here, each probability for the outcome is $\frac{1}{6}$. For a different dice with eight numbers, each probability would be $\frac{1}{8}$, but there would also be eight outcomes (Fig. 5.6).

In contrast, with the uniform probability distribution the probabilities of the lower outcomes are added together. As shown Fig. 5.5 for every higher outcome $\frac{1}{6}$ gets added. Also, the probability of $\mathbb{P}(A)$ is found in the outcome of four in the probability distribution, since A is the sum of all outcomes from 1 to 4 (see Example 5.3.2). This is also the definition of the probability distribution for $X = 4$ (see 5.7).

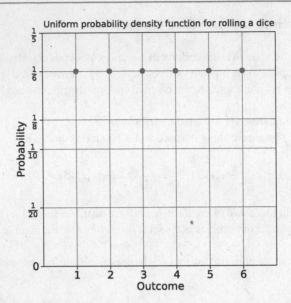

Figure 5.5 Density function of the discrete uniform distribution for rolling a dice

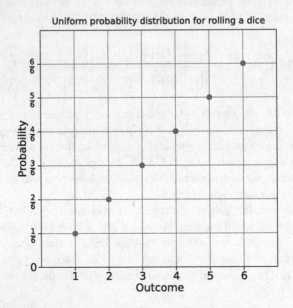

Figure 5.6 Distribution of the discrete uniform distribution for rolling a dice

5.3.3 Continuous Probability Distribution

While the discrete uniform distribution is a discrete distribution with a discrete sample set, a continuous distribution has a continuous sample set. These continuous sample sets have an infinite number of set elements. Therefore their probability distribution is calculated differently.

Definition 5.3.7: Continuous probability distribution
Let $(\Omega, \mathcal{P}(\Omega), \mathbb{P})$ be a probability space. Let X be a random variable with a probability density function $f : \mathbb{R} \to [0, 1]$ and $A = \{X \leq x\}$ an outcome. Then

$$F(A) = \mathbb{P}(X \leq x) = \int_{-\infty}^{x} \rho(t)dt = \int_{\Omega} \mathbb{1}_A(X)d\mathbb{P} \qquad (5.10)$$

is the continuous probability distribution of X (Henze 1997, Bovier 2019).

The term $\mathbb{1}_A$ is an indicator function and is defined by:

Definition 5.3.8: Indicator function
Let $\omega \in \Omega$ a element in the set Ω. Then

$$\mathbb{1}_A(\omega) = \begin{cases} 0, & \text{if } \omega \notin A \\ 1, & \text{if } \omega \in A \end{cases} \qquad (5.11)$$

the indicator function of ω for the set A (Henze 1997).

By using the indicator function, one can ensure that only the results of A are taken into account for the probability in equation 5.10. Every other element in Ω that is not contained in A is assigned a value of 0 in the integral. Therefore, only elements in A contribute to the result of the integral. The term $d\mathbb{P}$ therefore only replaces the integral across $\rho(t)dt$.

Example 5.3.4: Continuous uniform distribution with *spin the bottle*
Spin the bottle is a game where one spins a bottle in a circle in order to determine who has to do a dare. To describe this game mathematically, one can interpret the rotation of the bottle as a random variable. This random variable describes in which direction the bottle points once it has stopped spinning. This is determined by the extent to which the bottle rotates. To simplify this model, only the degrees between $0°$ and $360°$ are considered.

The event set of $[0°, 360°] \in \mathbb{R}^+$ is a set with an infinite number of events in which each event has the same probability. Therefore, this game can be described with the help of a continuous uniform distributed random variable, which would not be possible when using a discrete uniform distribution. This results in a density function as in Fig. 5.7:

$$\rho(t) = \frac{1}{360}.$$

Using equation 5.10, the probability distribution for any rotation between 0 and x degrees with $A = [0, x]$ is given by:

$$\mathbb{P}(A) = \int_\Omega \mathbb{1}_A(X) d\mathbb{P} = \int_0^x \rho(t) dt = \int_0^x \frac{1}{360} dt = \left[\frac{1}{360}t\right]_{t=0}^{t=x} = \frac{1}{360}x - 0 = \frac{1}{360}x.$$

Therefore, the probability distribution is a linear function as in Fig. 5.8. For example, if a person sits between 0° and 30° of the rotation, the probability of having to do a dare would be $\frac{1}{360} \cdot 30 = \frac{1}{12}$.

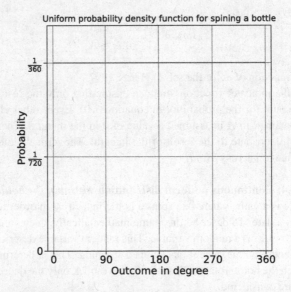

Figure 5.7 Density function of the continuous uniform distribution for spin the bottle

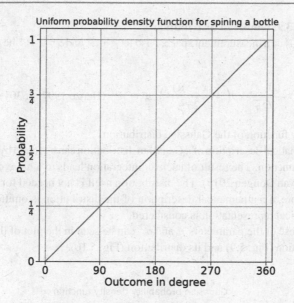

Figure 5.8 Distribution of the continuous uniform distribution for spin the bottle

In general, the density function for a given uniform distribution is defined by the upper limit b and lower limit a of the outcomes:

$$\rho(t) = \frac{1}{b - a} \text{ (Behrends 2012).} \tag{5.12}$$

5.3.4 Gaussian Distribution

The uniform distribution can simulate a indicator rating that does not have an attractor. There is no rating with a higher probability since every rating has exactly the same probability. Thus, there cannot be a rating that is favoured by the indicator. One way to apply higher probabilities for a given rating is to use the Gaussian distribution.

Definition 5.3.9: Gaussian distribution

Let $(\Omega, \mathcal{P}(\Omega))$ be a measurement space. Also let $\mu \in \mathbb{R}$ and $\sigma^2 \in \mathbb{R}^+$ be parameters. Then

$$\rho(x) = \text{``}\frac{1}{\sigma\sqrt{2\pi}}exp\left(-\frac{(x-\mu)^2}{2\sigma^2}\right), \ x \in \mathbb{R}\text{''} \quad \text{(Henze 1997, p. 301)} \qquad (5.13)$$

is the density function of the Gaussian distribution.

A mathematical description of a Gaussian distribution can be found by integrating the density function. The result of such an integration leads to a more complicated calculation (van Dongen 2015). The distribution itself is not needed for the simulation. Therefore, a mathematical description of the distribution is omitted here and only a graphical representation is considered.

The purpose of the parameters μ and σ^2 can be seen in the plot of the Gaussian density function (Fig. 5.9) and its distribution (Fig. 5.10).

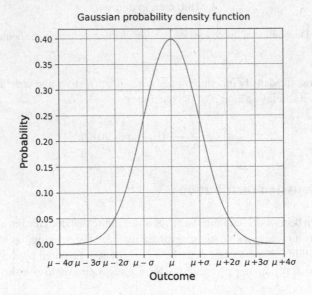

Figure 5.9 Density function of the continuous Gaussian distribution (ibid.)

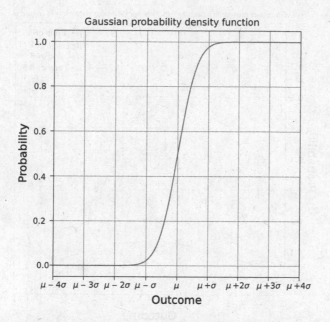

Figure 5.10 Distribution of the continuous Gaussian distribution (ibid.)

One can see from the Figures 5.9 and 5.10 that the result with the highest probability is μ. The probability of 50% in the distribution is at μ, since the density function is symmetrical. The results with a high impact in the distribution are between $\mu - \sigma$ and $\mu + \sigma$. Therefore, in the SSG-method, an indicator would have the most ratings in this interval and the most ratings at μ. The purpose of σ is explained in more detail in the following example.

Example 5.3.5: Gaussian density function with different values of σ
The Gaussian density function with fixed values for its parameters μ and σ can be seen in Fig. 5.11. The difference between the red and green curve is their value for μ. Higher μ values shift the density function to the left, while lower values shift the function to the right. In contrast, high values for σ compress the function and low values stretch the function (see blue and purple curve).

This means that for an indicator with a Gaussian distribution, higher values for σ decrease the variability of the rating. In contrast low values for σ increase the variability, as the probability for multiple different ratings increases.

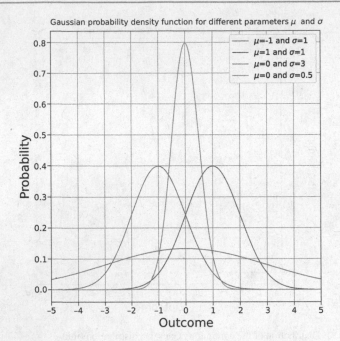

Figure 5.11 Density function of the continuous Gaussian distribution for different parameters

5.3.5   Brownian Motion

With the uniform distribution and the Gaussian distribution it is possible to distinguish between an indicator with a random rating and a indicator with different types of variability. A dynamic progression, where at each point the rating increases or decreases randomly depending on the previous score, is not possible with these distributions. The aim of this chapter is to formulate a random variable where at each point in time the outcome of the rating can randomly increase or decrease. The variability of this random increase or decrease must also be adjustable. With such a random variable, another type of variability can be simulated. This type of random variable, where a new random result is added at each time step, is called a stochastic process.

Definition 5.3.10: Stochastic process

Let $T \subset \mathbb{R}$ be a set. "A stochastic process is a set of random variables $\{X(t)|t \in T\}$ defined across the same probability space and indexed by the parameter t, called time. [...] The process is called continuous-time or discrete-time if the time parameter t is continuous or discrete, respectively" (Clark et al. 2007, p. 38). A stochastic process is called real valued if its random variables are real valued (Klenke 2020).

A simple realisation of a stochastic process is the one-dimensional random walk. In this case, the result at each step in time t can vary between an increase or a decrease by 1. This can be described mathematically by following example.

Example 5.3.6: One-dimensional simple random walk

Let $(Y_i)_{i \in T}$ and $T = \{1, 2, \ldots, n\}$. Also let Y_i be a random variable of the Bernoulli distribution which can be described as follows:

$$\text{"}\mathbb{P}(Y_i = 1) = 1 - \mathbb{P}(Y_i = -1) = p\text{"} \text{ (Bovier 2019, p. 56).} \tag{5.14}$$

Here S_n is the sum across all Y_i

$$\text{"}S_n = \sum_{i=1}^{n} Y_i\text{"} \text{ (ibid.)}$$

is the stochastic process. S_n is called the one-dimensional simple random walk on $\mathbb{Z}$ (Klenke 2020). Different realisations for S_n with $p = \frac{1}{2}$ can be seen in Fig. 5.12. Here each step in i from 1 to n corresponds to the time parameter t in the definition above. Since t is discrete the stochastic process S_n is discrete too.

As intended, at each step the result of S_n can either be increased by one or decreased by one. This leads to the random walks shown in Fig. 5.12. In contrast to the Gaussian distribution, where the variability is globally fixed, the variability of the random walk results from the different outcomes of the individual steps.

Instead of a random variable with just two outcomes (like -1 and 1 in the example above), a random variable with more outcomes can also be used. In the example above, the so called Bernoulli distribution (defined through equation 5.14) is used for each added random variable (Bovier 2019). A Gaussian distribution can also be used instead of the Bernoulli distribution. Such a stochastic process is called Brownian motion.

Definition 5.3.11: Brownian motion

Let $(\Omega, \mathcal{P}(\Omega), \mathbb{P})$ be a probability space. "A real valued stochastic process $\{B(t) : t \geq 0\}$ is called a (linear) Brownian motion with start in $x \in \mathbb{R}$ if the following holds:

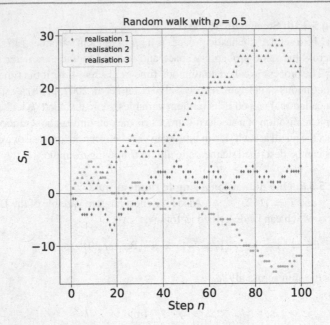

Figure 5.12 Different realisations for the random walk with p = 0.5

- $B(0) = x$,
- the process has independent increments, i.e. for all times $0 \leq t_1 \leq t_2 \leq \ldots \leq t_n$ the increments $B(t_n) - B(t_{n-1})$, $B(t_{n-1}) - B(t_{n-2})$, ..., $B(t_2) - B(t_1))$ are independent random variables,
- for all $t \geq 0$ and $h > 0$, the increments $B(t + h) - B(t)$ are" (Mörters & Peres 2010, p. 7) Gaussian distributed with $\mu = 0$ and $\sigma^2 = h$,
- "almost surely, the function $t \rightarrow B(t)$ is continuous" (ibid.), i.e. $B(t)$ is continuous expect on a set $\Omega' \subset \Omega$ where $\mathbb{P}(\Omega') = 0$ (Klenke 2020).

In contrast to the one-dimensional simple random walk it is possible to set a start point in the Brownian motion. This is given by the first premise $B(0) = x$, because the start point lies where $t = 0$ is inserted into $B(t)$. While in random motion each step S_n consists of Bernoulli distributed random variables, the Brownian motion has the increments $B(t_i) - B(t_{i-1})$ which are Gaussian distributed random variables. The third ensures that each step $B(t + h) - B(t)$ is Gaussian distributed. This means that each step itself is a continuous random variable, since the a Gaussian distributed random variable is continuous. μ has to be zero because the increments would otherwise shift in a certain direction, which is undesirable. Since the Brownian motion jumps between the increments, the function $t \rightarrow B(t)$ is not continuous. The values, for t, for which this function is not continuous are only points between $B(t_i) - B(t_{i-1})$ and $B(t_{i+1}) - B(t_i)$. These points have a probability measurement of zero, since every set of points has a probability measurement of zero (Behrends 2012). Therefore, the function $t \rightarrow B(t)$ is almost surely continuous. Unlike $mu = 0$ the parameter $\sigma^2 = h$ can be chosen freely. The influence of σ on the Brownian motion can be seen in following example.

Example 5.3.7: Brownian motion with different values for σ
Let $B(t)$ be a Brownian motion with Gaussian distributed increments with the parameter σ^2. For the values $\sigma = 2$, $\sigma = 5$, and $\sigma = 10$ a realisation can be seen in Fig. 5.13. The starting point for each realisation is kept to $x = 0$.

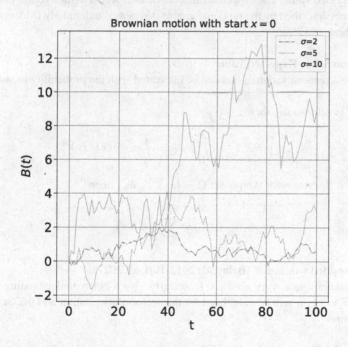

Figure 5.13 Brownian motion for different values of σ

The already mentioned non-continuous points of Brownian motion between the individual steps can be seen in Fig. 5.13. This is the case when the small Gaussian distributed pieces jump back and forth between the individual steps, whereby it is noticeable that at higher values of σ each step makes a larger jump.

Therefore, an indicator which follows a Brownian motion gains greater variability for higher values of σ. At the same time, the variability decreases for lower values of σ just as it does for a Gaussian distributed indicator.

5.4   Expected Value

In the ASSG-method, a value is needed that can measure a global value for the rating of the indicators. This would be a value that describes the average outcome of a random rating of a given indicator. A parameter that fulfils this sought-after property is the expected value. The expected value can be interpreted as the average outcome if the event described by the random variable is repeated indefinitely (Montgomery & Adelbratt 1982).

Definition 5.4.1: Expected value
Let X be a random variable, that can be integrated with the probability measure $\mathbb{P}$.

1. If X is continuous, then

$$\mathbb{E}(X) = \int_\Omega X d\mathbb{P} = \int_\Omega x \cdot \rho(x) dx \qquad (5.15)$$

2. If X is discrete with sample set $\Omega = \{\omega_1, \ldots, \omega_n\}$, then

$$\mathbb{E}(X) = \sum_{i=1}^{n} \omega_i \mathbb{P}(\omega_i) \qquad (5.16)$$

is the expected value of X (Behrends 2012, Birkner 2021).

This definition is very abstract. In order to give a better understanding of the expected value, it will be calculated for the uniform distribution and the Gaussian distribution.

Lemma 5.4.1: Expected value of the continuous uniform distribution
Let X be a continuous uniform distributed random variable with lower limit a and upper limit b, then the expected value of X is

$$E(X) = \frac{a+b}{2} \text{ (Behrends 2012).} \tag{5.17}$$

Proof 5.4.1: Expected value of the continuous uniform distribution
To prove the above lemma, the definition of the expected value is used. Since the density function of the uniform distribution is already defined by 5.12, the density function can be inserted into the definition. What then remains is the calculation of the integral of the Gaussian distributed random variable X.

$$E(X) = \int_\Omega x\rho(x)dx = \int_a^b \frac{1}{b-a}xdx = \frac{1}{b-a}\int_a^b xdx$$
$$= \frac{1}{b-a}\frac{x^2}{2}\Big|_a^b = \frac{b^2-a^2}{2(b-a)} = \frac{a+b}{2} \text{ (ibid.)}$$

$\square$

To put this newly found expected value into context, the example of rolling a *spin the bottle* is revisited.

Example 5.4.1: Expected value of *spin the bottle*
The highest possible outcome of the game is $b = 360°$ and the lowest outcome is $a = 0°$. Using the above lemma, this leads to the expected value $E(X) = 180°$. Since each outcome has the same probability, the average outcome is the middle value of all outcomes $E(X) = 180°$.

Lemma 5.4.2: Expected value of the discrete uniform distribution
Let X be a discrete uniform distributed random variable with lower limit $a = 1$ and upper limit $b = n$, then the expected value of X is

$$E(X) = \frac{n+1}{2} \text{ (Çalik \& Güngör 2004).} \tag{5.18}$$

Proof 5.4.2: Expected value of the discrete uniform distribution
Similar to the last proof, the definition is applied and the characteristics of the distribution are used. In this case, one can assume that the probability of each outcome is equal.

Let X be a discrete uniformly distributed random variable with the lower limit $a = 1$ and upper limit $b = n$. Then the probability of each outcome x_i with $i \in \{1, \ldots, n\}$ is:

$$\mathbb{P}(x_i) = \frac{1}{n} \text{ (Heine 2015).}$$

The Gaussian summation

$$\sum_{i=1}^{n} i = \frac{n(n+1)}{2} \text{ (Schindler-Tschirner \& Schindler 2019)}$$

is also used. With these two identities, the expected value can be calculated.

$$
\begin{aligned}
\mathbb{E}(X) = \sum_{i=1}^{n} \omega_i \mathbb{P}(\omega_i) &= \sum_{i=1}^{n} \omega_i \frac{1}{n} \\
&= \frac{1}{n} \sum_{i=1}^{n} \omega_i \\
&= \frac{1}{n} \sum_{i=1}^{n} i \\
&= \frac{1}{n} \frac{n(n+1)}{2} = \frac{n+1}{2} \text{(Çalik \& Güngör 2004).}
\end{aligned}
$$

$\square$

Example 5.4.2: Expected value of rolling a dice

Taking another look at the example of the random variable X with a dice roll gives the expected value:

$$\mathbb{E}(X) = \frac{n+1}{2} = \frac{6+1}{2} = 3.5 \text{ (Eichler \& Vogel 2011).} \tag{5.19}$$

To understand this example better, imagine a game in which one gets 1€ if a one is rolled and 2€ if a two is rolled and so on. Then the average win when the game is played would be 3.5€ (Montgomery & Adelbratt 1982).

As shown in the example above, the expected value is not the value of the outcome with the highest probability (ibid.). Nor does it have to be the actual outcome. For example, it is not possible to roll a 3.5. Instead, it is the average gain of one execution of the random variable.

However, it is possible that the average gain is equal to the outcome with the highest probability. One distribution where this is the case is the Gaussian distribution. In this distribution, the average gain is also the outcome with the highest probability μ.

Lemma 5.4.3: Expected value of the Gaussian distribution
Let X be a uniform distributed random variable, then the expected value of X is

$$\mathbb{E}(X) = \mu \text{ (Birkner 2021)}. \tag{5.20}$$

Proof 5.4.3: Expected value of the Gaussian distribution
The proof of this expected value is similar to the uniform distribution. In this case, the integral is solved by the following substitution:

$$z := \frac{x - \mu}{\sqrt{2}\sigma} \Leftrightarrow x = \sqrt{2}\sigma z + \mu, \text{ with } x, z, \mu \in \mathbb{R} \text{ and } \sigma \in \mathbb{R}_+.$$

This leads to the following integration constant of the substitution:

$$dz = \frac{1}{\sqrt{2}\sigma} dx \, (ibid.).$$

Let X be a Gaussian distributed random variable. Then the substitution leads to the following expected value:

$$
\begin{aligned}
\mathbb{E}(X) &= \int_{-\infty}^{\infty} x \cdot \frac{1}{\sigma\sqrt{2\pi}} exp\left(-\frac{(x-\mu)^2}{2\sigma^2}\right) dx \\
&= \frac{1}{\sqrt{\pi}} \int_{-\infty}^{\infty} (\sqrt{2}\sigma z + \mu) e^{-z^2} dz \\
&= \mu \cdot \frac{1}{\sqrt{\pi}} \int_{-\infty}^{\infty} e^{-z^2} dz + \frac{\sqrt{2}\sigma}{\sqrt{\pi}} \int_{-\infty}^{\infty} ze^{-z^2} dz \\
&= \mu \cdot \frac{1}{\sqrt{\pi}} \int_{-\infty}^{\infty} e^{-z^2} dz + \frac{\sqrt{2}\sigma}{\sqrt{\pi}} \lim_{c \to \infty} \left(-e^{-\frac{c^2}{2}} + e^{-\frac{c^2}{2}}\right) \\
&= \mu \cdot \frac{1}{\sqrt{\pi}} \int_{-\infty}^{\infty} e^{-z^2} dz + \frac{\sqrt{2}\sigma}{\sqrt{\pi}} \lim_{c \to \infty} 0 \\
&= \mu \cdot \frac{1}{\sqrt{\pi}} \int_{-\infty}^{\infty} e^{-z^2} dz \\
&\overset{*}{=} \mu \cdot 1 = \mu \text{ (ibid.).}
\end{aligned}
$$

$\square$

The equation ∗ is shown separately in following lemma.

Lemma 5.4.4: Equation ∗
Let $x \in \mathbb{R}$, then

$$\frac{1}{\sqrt{\pi}} \int_{-\infty}^{\infty} e^{-x^2} dx = 1 \text{ (ibid.)}. \tag{5.21}$$

Proof 5.4.4: Equation ∗
Equation 5.21 is obtained by squaring both sides. In addition, the following substitution with polar coordinates is made:

$$z^2 + y^2 = r^2 \text{ (Papula 2017)}.$$

The identity can be determined using the Pythagorean theorem (Maor 2019) in the triangle of Fig. 5.14.

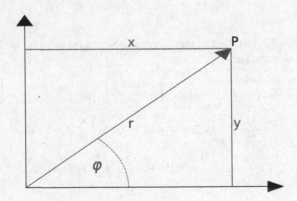

Figure 5.14 Relation between polar and Cartesian coordinates (Papula 2017, p. 42)

This leads to the integration constant of the substitution, a so called Jacobi determinant of the polar coordinates:

$$dydz = rdrd\varphi \text{ (Van Dongen 2015)}.$$

With this substitution and the following trick:

$$e^{-r^2} = \frac{d}{dr}\left(-\frac{1}{2}e^{-r^2}\right) \text{ (Birkner 2021)}$$

the integration can be solved as follows:

$$\left(\int_{-\infty}^{\infty} e^{-x^2}dx\right)^2 = \int_{-\infty}^{\infty}\int_{-\infty}^{\infty} e^{-(z^2+y^2)}dydz = \int_0^{2\pi}\int_0^{\infty} e^{-r^2} r\,dr\,d\varphi$$

$$= 2\pi\int_0^{\infty} e^{-r^2} r\,dr = 2\pi\int_0^{\infty} \frac{d}{dr}\left(-\frac{1}{2}e^{-r^2}\right)dr = -\pi e^{-r^2}\Big|_0^{\infty}$$

$$= -\pi \lim_{c\to\infty} e^{-c^2} + \pi e^{-0^2} = \pi \cdot 0 + \pi \cdot 1 = \pi \text{ (ibid.).}$$

If one rearranges this equation the following result can be obtained:

$$\left(\int_{-\infty}^{\infty} e^{-x^2}\right)^2 = \pi \Leftrightarrow \frac{1}{\sqrt{\pi}}\int_{-\infty}^{\infty} e^{-x^2}dx = 1$$

$$\square$$

5.5 Variance and Standard Deviation

With the expected value, a measure for a global rating of an indicator can be found. Another necessary value would be something that can estimate a global variability of an indicator. This can be done by the standard deviation. The standard deviation is directly obtained from the variance. "The variance is an intuitive measure of uncertainty" (Tamar et al. 2016, p. 664) that can be obtained by the average difference between the expected value and the outcomes of its random variable.

Definition 5.5.1: Variance
Let X be a random variable, that can be integrated with the probability measure $\mathbb{P}$ twice. Then

$$Var(X) = \mathbb{E}\left((x - \mathbb{E}(X))^2\right) \tag{5.22}$$

is the variance of X (Behrends 2012).

This definition is the same for continuous and discrete distributions, because it is simply a combination of the expected value (Eichler & Vogel). The standard deviation the square root of the variance. The advantage of a root variance will be explained in the first example.

Definition 5.5.2: Standard deviation
Let X be a random variable, that can be integrated with the probability measure $\mathbb{P}$ twice. Also let X have a non-negative variance $Var(X)$, then

$$\sigma(X) = \sqrt{Var(X)} \tag{5.23}$$

is the standard deviation of the random variable X (Eichler 2011).

Sometimes it is easier to calculate the expected value of $\mathbb{E}(X^2)$ instead of $\mathbb{E}\big((x - \mathbb{E}(X))^2\big)$. The following lemma can be used to calculate With $Var(X)$ with only $\mathbb{E}(X^2)$ and $\mathbb{E}(X)$. This will be convenient when calculating the variance of uniformly distributed random variables.

Lemma 5.5.1: Alternative formulation for $Var(X)$
Let X be a random variable, that can be integrated with the probability measure $\mathbb{P}$ twice. Then

$$Var(X) = \mathbb{E}(x^2) - \big(\mathbb{E}(X)\big)^2 \tag{5.24}$$

is the variance of X (Behrends 2012).

Proof 5.5.1: Alternative formulation for $Var(X)$
Let X be a random variable that can be integrated with the probability measure $\mathbb{P}$ twice. $Var(X)$ can thus be defined by equation 5.22. Since the expected value $\mathbb{E}(X)$ is an integral (see eq. 5.15), constants can be drawn in front of the integral. The integral of a sum can also be divided into sums of integrals. $\mathbb{E}(X)$ is a constant value, therefore it can also be drawn in front of an integral.

A second identity used in this proof is that the integration over the entire space of the probability density function is one (Klenke 2020). This is the case because the probability of obtaining any outcome of the sample space is 100%. Therefore the integration over the entire space of the probability density function, which is defined by the distribution function of the sample space has to be 1.

With these two arguments, the definition of $Var(X)$ can be rearranged as follows:

$$
\begin{aligned}
Var(X) &= \mathbb{E}\left(\left(x - \mathbb{E}(X)\right)^2\right) \\
&= \mathbb{E}\left(x^2 - 2\mathbb{E}(X)x + \left(\mathbb{E}(X)\right)^2\right) \\
&= \int \left(x^2 - 2\mathbb{E}(X)x + \left(\mathbb{E}(X)\right)^2\right)\rho(x)dx \\
&= \int x^2\rho(x)dx - \int 2\mathbb{E}(X)x\rho(x)dx + \int \left(\mathbb{E}(X)\right)^2\rho(x)dx \\
&= \mathbb{E}(x^2) - 2\mathbb{E}(X)\int x\rho(x)dx + \left(\mathbb{E}(X)\right)^2\int \rho(x)dx \\
&= \mathbb{E}(x^2) - 2\mathbb{E}(X)\mathbb{E}(X) + \left(\mathbb{E}(X)\right)^2\cdot 1 \\
&= \mathbb{E}(x^2) - \left(\mathbb{E}(X)\right)^2 \text{ (Behrends 2012).}
\end{aligned}
$$

$\square$

The proof for a discrete random variable is analogous to this proof. The only difference is that the expected value is a sum instead of an integral (ibid.).

With this lemma the variance of the discrete uniform distribution can be calculated.

Lemma 5.5.2: Variance of the discrete uniform distribution
Let X be a discrete uniformly distributed random variable with the lower limit $a = 1$ and upper limit $b = n$, then

$$
Var(X) = \frac{n^2 - 1}{12} \text{ (Sandelius 1967).} \tag{5.25}
$$

Proof 5.4.2: Variance of the discrete uniform distribution
Let X be a discrete uniformly distributed random variable with the lower limit $a = 1$ and upper limit $b = n$. Similar to the proof of the expected value, an identity of a sum is needed:

$$
\sum_{i=1}^{n} i^2 = \frac{n(n + 1)(2n + 1)}{6} \text{ (Schindler-Tschirner \& Schindler 2019).}
$$

With the alternative formulation of $Var(X)$ only $\mathbb{E}(X)$ and $\mathbb{E}(X^2)$ is need. $\mathbb{E}(X)$ was already calculated for the random variable X in eq. 5.18. This only leaves $\mathbb{E}(X^2)$ to be calculated separately here with the above sum identity form.

$$\mathbb{E}(X^2) = \sum_1^n x^2 \mathbb{P}(X)$$

$$= \sum_{i=1}^n x^2 \frac{1}{n}$$

$$= \frac{1}{n} \sum_1^n x^2$$

$$= \frac{1}{n}\left(1^2 + 2^2 + 3^2 + \ldots + n^2\right)$$

$$= \frac{1}{n}\left(\frac{n(n+1)(2n+1)}{6}\right)$$

$$= \frac{(n+1)(2n+1)}{6} \tag{5.26}$$

Using this newly found identity and the already calculated $\mathbb{E}(X)$ the calculation of eq. 5.24 remains:

$$Var(X) = \mathbb{E}(x^2) - \left(\mathbb{E}(X)\right)^2$$

$$= \frac{(n+1)(2n+1)}{6} - \left(\frac{n+1}{2}\right)^2$$

$$= \frac{2n^2 + 3n + 1}{6} - \left(\frac{n+1}{2}\right)^2$$

$$= \frac{2n^2 + 3n + 1}{6} - \frac{n^2 + 2n + 1}{4}$$

$$= \frac{4n^2 + 6n + 2}{12} - \frac{3n^2 + 6n + 3}{12}$$

$$= \frac{4n^2 - 3n^2 + 6n - 6n + 2 - 3}{12}$$

$$= \frac{n^2 - 1}{12} \text{ (Sandelius 1967).}$$

□

Example 5.5.1: Variance of rolling a dice

Now the example of a discrete, uniformly distributed random variable X of rolling a dice can be taken up again. With the newly found formula for the variance the variance of X using $a = 1$ and $n = 6$ is

$$Var(X) = \frac{n^2 - 1}{12} = \frac{6^2 - 1}{12} = \frac{35}{12} \approx 2.62$$

In the game described in the last example, the variance would be $2.62 \, \text{€}^2$. However, a value in €^2 can not be put into context with a value in €. This is where the standard deviation of the random variable of this game, which will be called X', can be helpful:

$$\sigma(X') = \sqrt{Var(X')} \approx 1.71\text{€}.$$

With the standard deviation, the variability of the game can be put into context. This means that the variability of the average gain of one play of the game is 1.71 €. Whether this value is high or low can only be estimated in comparison to other standard deviations.

As with the expected value, the variance of the continuous uniform distribution is also different from the variance of the discrete uniform distribution.

Lemma 5.5.3: Variance of the continuous uniform distribution
Let X be a discrete uniformly distributed random variable with the lower limit a and the upper limit b, then

$$Var(X) = \frac{(b - a)^2}{12} \text{ (Behrends 2012).} \qquad (5.27)$$

Proof 5.4.3: Variance of the continuous uniform distribution
Let X be a discrete uniformly distributed random variable with the lower limit a and the upper limit b. As with the proof of the discrete version, the alternative formula for the variance is also used for the continuous version. Here, the expected value $\mathbb{E}(X)$ from equation 5.17 can be used.

A second identity that is needed is:

$$\begin{aligned}
(b - a)^3 &= (b - a)^2(b - a) \\
&= (b^2 - 2ab + a^2)(b - a) \\
&= b^3 - 2ab^2 + a^2b - b^2a + 2a^2b + a^3 \\
&= b^3 - 2ab^2 - ab^2 + 2a^2b + a^2b + a^3 \\
&= b^3 - 3ab^2 + 3a^2b + a^3.
\end{aligned}$$

With these both identities the variance is:

$$Var(X) = \mathbb{E}(x^2) - \left(\mathbb{E}(X)\right)^2 = \int x^2 d\mathbb{P} - \left(\frac{a+b}{2}\right)^2$$

$$= \int_a^b \frac{x^2}{b-a} dx - \left(\frac{a+b}{2}\right)^2 = \frac{x^3}{3(b-a)}\Big|_a^b - \left(\frac{a+b}{2}\right)^2$$

$$= \frac{b^3 - a^3}{3(b-a)} - \frac{(a+b)^2}{4}$$

$$= \frac{1}{12(b-a)}\left(4b^3 - 4a^3 - 3(a^2 + 2ab + b^3)(b-a)\right)$$

$$= \frac{1}{12(b-a)}\left(4b^3 - 4a^3 - 3a^2b - 3ab^2 + 3b^3 + 3a^3 + 6a^2b + 3ab^2\right)$$

$$= \frac{1}{12(b-a)}\left(4b^3 - 3b^3 + 3ab^2 - 6ab^2 - 3a^2b + 6a^2b - 4a^3 + 3a^3\right)$$

$$= \frac{1}{12(b-a)}\left(b^3 - 3ab^2 + 3a^2b - a^3\right)$$

$$= \frac{1}{12(b-a)}(b-a)^3 = \frac{(b-a)^2}{12} \text{ (ibid.).}$$

$\square$

While μ was the expected value of the Gaussian distribution, it turns out that the variance of the Gaussian distribution is σ^2. This also means that the standard deviation of a Gaussian distribution is σ, which is why σ was used as a symbol for this parameter.

Lemma 5.5.4: Variance of Gaussian distribution
Let X be a continuous Gaussian distributed random variable with the parameters μ and σ, then

$$Var(X) = \sigma^2 \text{ (Birkner 2021).} \qquad (5.28)$$

Proof 5.4.4: Variance of Gaussian distribution
Let X be a continuous Gaussian distributed random variable with the parameters μ and σ. In this proof, the original definition of $Var(X)$ will be used. The same substitution of the proof of the expected value of a Gaussian distributed random variable is also needed. This was:

$$z := \frac{x - \mu}{\sqrt{2}\sigma} \Leftrightarrow x = \sqrt{2}\sigma z + \mu, \text{ with } x, z, \mu \in \mathbb{R} \text{ and } \sigma \in \mathbb{R}_+.$$

and the integration constant of the substitution:

$$dz = \frac{1}{\sqrt{2}\sigma}dx \, (ibid.).$$

Additionally the identity

$$e^{-r^2} = \frac{d}{dr}\left(-\frac{1}{2}e^{-r^2}\right) \text{ (Birkner 2021)}$$

will be used again for a partial integration. This leads to the following integration:

$$
\begin{aligned}
Var(X) &= \mathbb{E}\left((x - \mathbb{E}(X))^2\right) \\
&= \int_{-\infty}^{\infty} (x - \mu)^2 \rho(x) dx \\
&= \int_{-\infty}^{\infty} (x - \mu)^2 \frac{1}{\sigma\sqrt{2\pi}} exp\left(-\frac{(x - \mu)^2}{2\sigma^2}\right) dx \\
&= \int_{-\infty}^{\infty} \sigma^2 z^2 \frac{1}{\sqrt{2\pi}} e^{-\frac{z^2}{2}} dz \\
&= \frac{\sigma^2}{\sqrt{2\pi}} \int_{-\infty}^{\infty} z^2 e^{-\frac{z^2}{2}} dz \\
&= \frac{\sigma^2}{\sqrt{2\pi}} \int_{-\infty}^{\infty} z\left(-\frac{d}{dz}e^{-\frac{z^2}{2}}\right) dz \\
&= \frac{\sigma^2}{\sqrt{2\pi}} \left(\left[z\left(-e^{-\frac{z^2}{2}}\right)\right]_{-\infty}^{\infty} - \int_{-\infty}^{\infty} 1 \cdot \left(-e^{\frac{z^2}{2}}\right) dz\right) \\
&= \frac{\sigma^2}{\sqrt{2\pi}} \left(\left[-ze^{-\frac{z^2}{2}}\right]_{-\infty}^{\infty} + \sqrt{2\pi} \cdot \frac{1}{\sqrt{2\pi}} \int_{-\infty}^{\infty} e^{\frac{z^2}{2}} dz\right) \\
&= \frac{\sigma^2}{\sqrt{2\pi}} \left(\lim_{c \to \infty}\left(-ce^{-\frac{c^2}{2}} + ce^{-\frac{c^2}{2}}\right) + \sqrt{2\pi} \cdot \frac{1}{\sqrt{2\pi}} \int_{-\infty}^{\infty} e^{\frac{z^2}{2}} dz\right) \\
&= \frac{\sigma^2}{\sqrt{2\pi}} (\lim_{c \to \infty} 0 + \sqrt{2\pi} \cdot 1) \\
&= \sigma^2 \, (ibid.).
\end{aligned}
$$

□

With these proven variances, the variability of every simulated indicator ranking following these distributions can be calculated. This gives a first insight into the global variability of the indicators in the ASSG-method.

5.6   Law of Large Numbers

In contrast to a simulated indicator rating the distribution of a "real" rating of an indicator in a students-teacher system is unknown. Therefore, the expected values and variances of the previous chapters cannot be used. This is where the law of large numbers can be used. With this theorem, the expected value of an unknown distribution can be approximated.

· The law of large numbers can be divided into different kinds of approximations. This leads to the weak law of large numbers and the strong law of large numbers. Since both types are needed for the ASSG-method and its simulation, both laws will be covered in this chapter.

5.6.1   Weak Law of Large Numbers

In order to better understand the idea of the law of large numbers, a simple example is given made.

Let X_i be a discrete uniform distributed random variable. Then $(X_i)_{i \in \{1,\dots,n\}}$ is a set of n discrete uniform distributed random variables. With the results of these random variables, a mean value is calculated. From this, the mean value $S(n)$ results as follows:

$$S(n) = \frac{1}{n} \sum_{i=1}^{n} X_i. \qquad (5.29)$$

$S(n)$ is the mean value of n rolled dice. This $S(n)$ can for instance be determined for a dice that is rolled $n = 10$ times. For example, the values of ten rolls could be $\{1, 2, 3, 6, 4, 2, 4, 6, 1, 3\}$. With these rolls $S(n = 10)$ would be equal to:

$$S(10) = \frac{1}{10}\left(1 + 2 + 3 + 6 + 4 + 2 + 4 + 6 + 1 + 3\right) = \frac{32}{10} = 3.2.$$

This process of rolling $n = 10$ dice and then calculating $S(10)$ is also done $n = 100$ times. The results are then transferred to a histogram. A simulated result of such a histogram can be seen in the upper left plot of Fig. 5.15.

If one were to make twenty throws for a mean value instead of ten, this would lead to the representation at the top right. The same applies to $n = 100$ and $n = 10000$ for the last two plots.

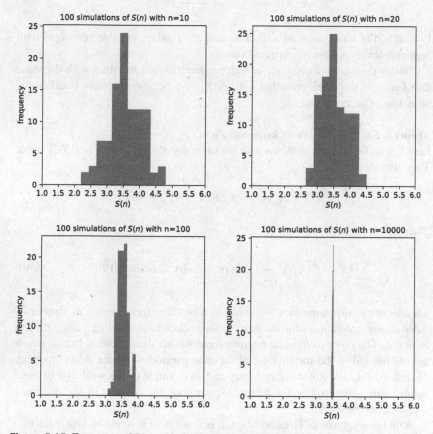

Figure 5.15 Frequency of S(n) with fixed values of n and 100 simulations of S(n)

As can be seen in the histograms, the largest mass is at 3.5. This is already the case for $n = 10$, but for higher values the histogram gets even narrower. This process eventually gets to the point where virtually all of the results for $S(n)$ are

3.5 at $n = 10000$. This type of convergence is called the convergence of probability (Klenke 2020).

The value to which $S(n)$ converges is the expected value of the random variable X_i. Since X_i is the random variable of a discrete uniform distribution with lower limit 1 and upper limit $n = 6$ the formula of 5.18 applies:

$$\mathbb{E}(X_i) = \frac{n+1}{2} = \frac{6+1}{2} = 3.5.$$

Evidently, the mean value of a discrete uniform random variable converges with high probability against its expected value.

This is the case for every set of independent random variables with the same distribution if the distribution has a $Var(|X_i|) < \infty$. This theorem is called the weak law of large numbers.

Theorem 5.6.1: Weak law of large numbers
Let $(X_i)_{i \in \mathbb{N}}$ be random variables with the same distribution with $Var(|X_i|) < \infty$. They also satisfy:

$$\mathbb{E}(X_i X_j) = \mathbb{E}(X_i)\mathbb{E}(X_j) \, \forall i, j \in \mathbb{N}.$$

Then:

$$\frac{1}{n} \sum_{i=1}^{n}(X_i - \mathbb{E}(X_i)) \stackrel{n \to \infty}{\to} 0 \text{ in probability (Bovier 2019)}. \tag{5.30}$$

The electronic supplementary material has a proof of this theorem in chapter A. Additionally a definition for the term "convergence in probability" can be found there too. The proof of this law requires some further mathematical basics, which are not needed in the formulation of the new parameters of the ASSG method. Therefore, only the following corollary and an example of the weak law of large numbers are given in this. For further explanations see chapters of the electronic supplementary material.

With the weak law of large numbers, it is now possible to get an approximation of the expected value using the mean value. Attention only has to be paid to the fact that the number of samples for the mean value is large enough. If this is the case, the following approximation can be proven.

Corollary 1.6.1: Approximation of $\mathbb{E}(X)$ with the mean value

Let $(X_i)_{i \in \{1,\dots,n\}}$ be a set of random variables that satisfy the weak law of large numbers. Then the following applies for large enough n:

$$\mathbb{E}(X_1) \approx \frac{1}{n} \sum_{i=1}^{n} X_i \text{ (van Dongen 2015).} \tag{5.31}$$

Proof 5.6.1: Approximation of $\mathbb{E}(X)$ with the mean value

Let $(X_i)_{i \in \{1,\dots,n\}}$ be a set of random variables that satisfy the weak law of large numbers. If n is large enough, then using the weak law of large numbers, then the following applies:

$$\frac{1}{n} \sum_{i=1}^{n} (X_i - \mathbb{E}(X_i)) \approx 0.$$

Since all X_i are identically distributed, their expected values are also the same, so that it is irrelevant which expected value is calculated, because e.g. $\mathbb{E}(X_i) = \mathbb{E}(X_1) \ \forall i \in \{1, \dots, n\}$ is true. With this argument, the above sum can be rearranged:

$$0 \approx \frac{1}{n} \sum_{i=1}^{n} (X_i - \mathbb{E}(X_i)) = \frac{1}{n} \sum_{i=1}^{n} (X_i - \mathbb{E}(X_1)) = \frac{1}{n} \left(\sum_{i=1}^{n} (X_i) - n\mathbb{E}(X_1) \right)$$

$$= \frac{1}{n} \sum_{i=1}^{n} X_i - \frac{1}{n} n\mathbb{E}(X_1) \Leftrightarrow \frac{1}{n} \sum_{i=1}^{n} X_i \approx \mathbb{E}(X_1).$$

$\square$

Example 5.6.1: Weak law of large numbers and rolled dice

Let $(X_i)_{i \in \{1,\dots,n\}}$ be a set of discrete uniformly distributed random variables with upper limit $b = 6$ and lower limit $a = 1$. This is a set of random variables that describe n rolled dice. Since every X_i has the same distribution, they are identically distributed. Therefore, they satisfy the first premise of the weak law of large numbers.

The second premise can be proven with eq. 5.25:

$$Var(X) = \frac{n^2 - 1}{12}.$$

This variance is positive for any possible value of $n > 1$. $n = 1$ would only describe one throw of the dice. For a single throw, a mean value would not make sense. Therefore, it is possible to set $n > 1$ for this example. This leads to the second premise of the weak law of large numbers:

$$Var(|X_i|) = Var(X_i) = \frac{n^2 - 1}{12} = \frac{6^2 - 1}{12} = \frac{35}{12} < \infty. \tag{5.32}$$

The only thing left is the last premise. Since the outcome of one throw has no influence on the outcome of another throw, they are independent. If the random variables X_i are independent, their expected value can be calculated independently (Klenke 2020). Thus, they also fulfil all the premises of the weak law of large numbers and the mean converges to the expected value of X_i. For this reason, the convergence in the probability calculation can be seen in Fig. 5.15.

5.6.2 Strong Law of Large Numbers

The weak law of large numbers describes a convergence of the mass in a histogram such as in Fig. 5.15. By contrast, the strong law of large numbers describes a convergence of the function $S(n)$ at higher values of n. This can be illustrated in Fig. 5.17 with the same example of rolling n dice (Fig. 5.16).

This type of convergence is called an "almost sure" convergence. This convergence implies that if one just waits long enough (i.e. make enough rolls), the mean will be equal to the expected value. That is always the case unless one is extremely unlucky. Extremely unlucky in this case means, in mathematical terms, that an outcome with the probability 0 occurs. Such convergence is needed when results for parameters are simulated in the ASSG method.

This is a different kind of convergence than the convergence in probability. What is meant by the weak law of large numbers is that the errors that occur in the approximation between the expected value and the mean are small if n is large enough. However, the probability of making an error is never zero. Therefore, the strong law of large numbers is a stronger statement than the weak law of large numbers. This also means that the conditions for the convergence of a set of random variables are stronger for the strong law of large numbers than for the weak law of large numbers.

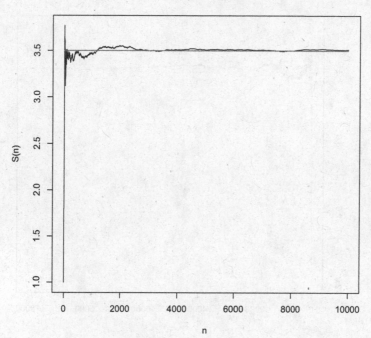

Figure 5.16 S(n) in dependence of n

Theorem 5.6.2: Strong law of large numbers

Let $(X_i)_{i \in \mathbb{N}}$ be identically distributed and independent random variables with $\mathbb{E}(X_1^4) < \infty$. Then they satisfy:

$$\frac{1}{n} \sum_{i=1}^{n} \left(X_i - \mathbb{E}(X_i) \right) \overset{n \to \infty}{\to} 0 \text{ almost surely (Bovier 2019).}$$

As with the weak law of large numbers, the proof of this theorem and the mathematical definition of almost sure convergence can be found in the electronic supplementary material chapter A.

A connection between the strong and the weak law of large numbers can be explained in the following diagram.

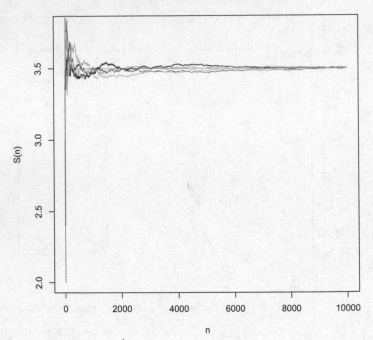

Figure 5.17 S(n) Different realisations of S(n) in dependence of n

Fig. 5.17 shows different realisations of $S(n)$. If one inserts the values of each $S(n)$ for the fixed value of $n = 100$ into a histogram, one obtains a histogram like the one in Fig. 5.15. This can be done for various values of n. Thus one can obtain different histograms like in the weak law of large numbers in last chapter. Since the curves in Fig. 5.17 converge towards the expected value for higher values of n, the histograms also become narrower for higher values of n. Therefore, one obtains the same kind of convergence in the histograms as with the weak law of large numbers.

How the premises of the strong law of large numbers can be satisfied is shown in this example:

Example 5.6.2: Strong law of large numbers and rolled dice
Let $(X_i)_{i \in \{1,...,n\}}$ be a set of discrete uniform distributed random variables with upper limit $b = 6$ and lower limit $a = 1$. This is a set of random variables that

describe n rolled dice. Since each X_i has the same distribution, they are identically distributed. Since the result of one roll has no influence on the result of another roll, they are independent of each other. Therefore, they fulfil the first and second premises of the weak law of large numbers.

The last premise can be proven with this calculation:

$$\mathbb{E}(X_1^4) = \sum_{n=1}^{6} n^4 \mathbb{P}(X_1^4 = n^4) = \frac{1}{6} \sum_{n=1}^{6} n^4 < \infty.$$

Thus the random variables also fulfil all the premises of the strong law of large numbers and the mean converges almost certainly towards the expected value of X_i. This is the reason why the almost sure convergence can be seen in Fig. 5.17.

Parameters of the Advanced State Space Grids

6

With the help of the mathematical foundations of the last chapter, it is now possible to formulate the new parameters of the ASSG method. This answers the first research question "How can the State Space Grids method be enhanced to the Advanced State Space Grid method?". The following subchapters will serve as a successive introduction to the new parameters. Subsequently, the interpretation of these parameters and the ASSG plot will be discussed in Chapter 7. The first objective of the new parameters is an estimate for a global outcome of the indicators in the ASSG method. Global in this case means that the entire duration of the ratings is covered. Furthermore, the types of global dynamic progression of the indicators and their ratings will be categorised with additional parameters. This is the first step towards answering the second research question "How can relation types be categorized in the Advanced State Space Grid method?".

For the definition of the parameters in this chapter the ASSG method will be kept to only two indicators on an x and y-axis. It is also possible to generalise these parameters to $m \in \mathbb{N}$ indicators in a single ASSG plot. However, with three indicators the ASSG plot would be three-dimensional and can therefore not be plotted easily in an two dimensional space. Moreover, a combination of more than three indicators would lead to a hyperspace in which the ASSG method would have to be plotted. A graphic illustration of such a hyperspace ASSG plot would be even more complicated and is not needed to find relationships between two aspects of the student-teacher system. Instead, the indicators can be combined to pairs of two dimensional ASSG plots. This would result in a vast number of ASSG plots, that could all be interpreted independently. Therefore, only combinations of two indicators in one ASSG plot are discussed in this chapter.

N. Litzenberger, *Introduction of Advanced State Space Grids and Their Application to the Analysis of Physics Teaching*, BestMasters,
https://doi.org/10.1007/978-3-658-42732-0_6

6.1   Grid and Ratings

First of all, the grid of the ASSG plot is defined. The following definition is limited to two indicators with the same number of possible rankings, as the indicators used for this method, will have the same ranking size.

Definition 6.1.1: Grid of an ASSG plot
Let the ASSG method only include two indicators with a possible ranking between 1 and $o \in \mathbb{N}$. Then the plot of an ASSG is an illustration of the vector space in $[1, o]x[1, o] \subset \mathbb{R}^+ x \mathbb{R}^+$. The grid of an ASSG plot is then a vector subspace in $\{1, ..., o\}x\{1, ..., o\} \subset \mathbb{N}x\mathbb{N}$.

This definition can easily be generalised for indicators with different rankingsizes. Two indicators with a maximum ranking of o and o' would result in a $[1, o]x[1, o']$ vector space.

The indicator itself is represented by random variables with an unknown distribution. However, the possible ratings and thus the results of these random variables can be defined as follows:

Definition 6.1.2: Rating of a single indicator
Let x be the random variable of an indicator on the x-axis of the ASSG plot. Then $\{x_1, x_2, x_n\}$ with $x_i \in \{1, ..., o\} \subset \mathbb{N}$ are the ratings of the indicator on the x-axis with the total number of n ratings. In this case, every rating x_i is a random variable and x is a set of the random variables $(x_i)_{i \in \{1, ..., n\}}$. Likewise the indicator on the y-axis is defined as y with the ratings $\{y_1, y_2, ..., y_n\}$.

Definition 6.1.3: Points in an ASSG
Let x and y be the indicators of the ASSG method with the number of ratings in both indicators n. Then every point in the ASSG is defined by $(x_i, y_i)^T$ for all $i \in \mathcal{I} = \{1, ..., n\} \subset \mathbb{N}$.

With the possible ratings of the indicators and the points in the ASSG the probability space of a single indicator can be described as follows:

Corollary 6.1.1: Probability space of the rating of a single indicator
Let $\Omega = \{1, 2, ..., n\}$ be the sample space of the random variable of one indicator in the ASSG with a probability measurement $\mathbb{P}$. Then the probability space of a ranking $(x_i)_{i \in \{1, ..., n\}}$ of one indicator x is $(\{1, 2, ..., n\}, \mathcal{P}(\{1, 2, ..., n\}), \mathbb{P})$.

Proof 6.1.1: Probability space of the rating of a single indicator
This corollary can be proven with the definition of a probability space (see eq. 5.7). Let $\Omega = \{1, 2, ..., n\}$ be the sample space of the random variable of one indicator in the ASSG with a probability measurement $\mathbb{P}$. Then insert $\{1, 2, ..., n\}$ for Ω and the corollary is proven. $\qquad\qquad\qquad\qquad\qquad\qquad\qquad\qquad\qquad\qquad\qquad\qquad\quad$ $\square$

This probability space can be generalized for a two dimensional ASSG method with a new sample space Ω. This sample space is a two dimensional combination of the sample spaces of two indicators.

Corollary 6.1.2: Probability space for one point in the ASSG
Let $\Omega_x = \{1, 2, ..., n\}$ be the sample space of the random variable of one rating x_i of the indicator on the x-axis in the ASSG with a probability measurement $\mathbb{P}_x$. Likewise let $\Omega_y = \{1, 2, ..., n\}$ and $\mathbb{P}_y$ be the corresponding terms for the rating y_i of the indicator on the y-axis. Then the probability space for the ASSG of both indicators is:

$$(\{1, 2, ..., n\}x\{1, 2, ..., n\}, \mathcal{P}(\{1, 2, ..., n\}x\{1, 2, ..., n\}), \mathbb{P}), \qquad (6.1)$$

with a new probability measurement $\mathbb{P}$.

Proof 6.1.2: Probability space for one point in the ASSG
The proof for this corollary is analogous to the previous proof. In this case Ω is $\{1, 2, ..., n\}x\{1, 2, ..., n\}$. $\qquad\qquad\qquad\qquad\qquad\qquad\qquad\qquad\qquad\qquad\qquad\qquad\quad$ $\square$

The new probability measurement $\mathbb{P}$ in this corollary can be described with $\mathbb{P}_x$ and $\mathbb{P}_y$ (Bovier 2019). However, since $\mathbb{P}_x$ and $\mathbb{P}_y$ are already unknown, there is no benefit in knowing more about $\mathbb{P}$ if it depends on $\mathbb{P}_x$ and $\mathbb{P}_y$.

Based on these definitions and probability distributions the ratings can be simulated and plotted as in Fig. 6.1.

This plot was created with the simulation in Chapter 8 and will be discussed in detail in that chapter. For now, these simulated ASSG plots only serve as examples for this chapter. A content analysis of the ASSG plots is not carried out for these examples. This is down to the fact that there are no real lessons for these examples. A content analysis would therefore not be meaningful. Such interpretation will be done with actual indicators in a student-teacher system in Chapter 9 and Chapter 10. In this and the following chapter, the interpretation will be limited to the theoretical analysis of the position of the points in the ASSG plot.

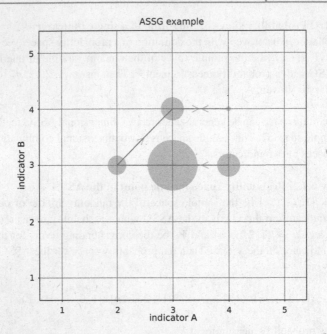

Figure 6.1 ASSG with grid and rating of its indicators

The circle diameter at each point of the grid is defined by the total time that the ratings of both indicators remain at that point. The following definition can be used to describe this circle diameter. Here $\mathbb{1}$ is the indicator function of eq. 5.11:

Definition 6.1.4: Circle diameter and dwell time on one location in the grid
Let x and y be the indicators of the ASSG method with n ratings. Then the size of the circle diameter s_i of the location of the point $(x_i, y_i)^T$ is equal to

$$s_i = \sum_{j=1}^{n} \mathbb{1}_{\{j \in \mathbb{N} | (x_i, y_i)^T = (x_j, y_j)^T\}}(j). \tag{6.2}$$

If the time between each rating remains the same and the total duration of the ratings is T minutes, then the total time that a combination of ratings $(x_i, y_i)^T$ is fulfilled is equal to the dwell time at that location in the grid:

$$t_i = \frac{T}{n} \cdot s_i. \tag{6.3}$$

In this definition of s_i, each point before the i-th point up to the last point is checked with the indicator function. Each time the value of $(x_i, y_i)^T$ equals the value of the j-th point in the sum the indicator function is 1. If the value of the i-point is not equal to the value of the j-th point, then the indicator function is 0. This results to sums of zeros and ones for s_i. Therefore s_i is equal to the number of encounters with the point in the grid during the entire duration of the rating. Since this is very theoretical an example will be made.

Example 6.1.1: Circle diameter and dwell time on one location in the grid
IImagine two indicators that are rated nine times in a lesson of 45 minutes in total. Let the ratings for indicator x be

$$\{x_1, x_2, x_3, x_4, x_5, x_6, x_7, x_8, x_9\} = \{1, 2, 3, 2, 2, 5, 3, 4, 2\}.$$

Also let the ratings of y be

$$\{y_1, y_2, y_3, y_4, y_5, y_6, y_7, y_8, y_9\} = \{1, 2, 5, 2, 2, 3, 2, 1, 2\}.$$

Then the points of the ASSG plot would be

$$\{(x_1, y_1)^T, (x_2, y_2)^T, (x_3, y_3)^T, (x_4, y_4)^T, (x_5, y_5)^T,$$
$$(x_6, y_6)^T, (x_7, y_7)^T, (x_8, y_8)^T, (x_9, y_9)^T\}$$
$$=\{(1, 1)^T, (2, 2)^T, (3, 5)^T, (2, 2)^T, (2, 2)^T, (5, 3)^T, (3, 2)^T, (4, 1)^T, (2, 2)^T\}.$$

With these points of the plot, the circle diameter of the fifth point can be calculated as follows:

$$s_5 = \sum_{j=1}^{9} \mathbb{1}_{\{j \in N | (2,2)^T - (x_j, y_j)^T\}}(j)$$

$$= \mathbb{1}_{\{j \in N | (2,2)^T = (1,1)^T\}}(j = 1) + \mathbb{1}_{\{j \in N | (2,2)^T = (2,2)^T\}}(j = 2)$$

$$+ \mathbb{1}_{\{j \in N | (2,2)^T = (3,5)^T\}}(j = 3) + \mathbb{1}_{\{j \in N | (2,2)^T = (2,2)^T\}}(j = 4)$$

$$+ \mathbb{1}_{\{j \in N | (2,2)^T = (2,2)^T\}}(j = 5) + \mathbb{1}_{\{j \in N | (2,2)^T = (5,3)^T\}}(j = 6)$$

$$+ \mathbb{1}_{\{j \in N | (2,2)^T = (3,2)^T\}}(j = 7) + \mathbb{1}_{\{j \in N | (2,2)^T = (4,1)^T\}}(j = 8)$$

$$+ \mathbb{1}_{\{j \in N | (2,2)^T = (2,2)^T\}}(j = 9)$$

$$= 0 + 1 + 1 + 0 + 1 + 0 + 0 + 0 + 1 = 4$$

Hence the circle diameter of the fifth point is 4, which also corresponds to the number of times the same state was met in the whole duration of the time sequences.

The time the state of the fifth point remains in the student-teacher system would then be

$$t_5 = \frac{45\,min}{9} \cdot 4 = 20\,min$$

if the ratings are made at fixed equal time intervals. So the state $(2, 2)^T$ is met for a global duration of 20 min.

Another simplification which is made in this chapter is that every indicator is rated in fixed time intervals. It is possible to change these intervals throughout one rating. This would lead to a different definition of the size of the circle diameters, as the time samples are not dependent on the order of the x_i values. As there will be no rating of an indicator without fixed time intervals in any of the following indicators, a generalisation of the dwell time and size will not be needed.

To get a better understanding of where the first and last point of the ASSG plot is, these two points will be defined.

Definition 6.1.5: Start value and end value for ASSGs
Let x and y be the indicators of the ASSG method. Then the start value of the ASSG plot is

$$\text{start value} = (x_1, y_1)^T \tag{6.4}$$

and the end value is

$$\text{end value} = (x_n, y_n)^T. \tag{6.5}$$

The start value and end value of the ASSG will be highlighted by green and yellow dots. With the help of these two dots and the associated arrows between the dots, the dynamic course of the indicators can be followed over the time of their ratings.

As can be seen in Fig. 6.2, the points in the plot start in middle, then jump to the right side, return to the middle, stay there for a long period of time, then go to the left, stay there for a short time, go to $(4, 3)^T$, stay there shortly, jump to the right, jump back again, and finally stay in $(3, 4)^T$ until the end.

The plot is still not a clear progression because it is not known exactly how long the points stay in one position before they move again. It is also possible, for example, that the points jump around numerous times between $(3, 3)^T$ and $(4, 3)^T$

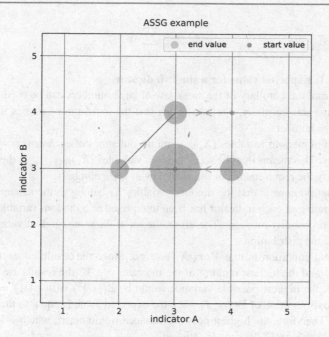

Figure 6.2 ASSG with start value and end value

before they go to $(2, 3)^T$. However, the aim of the ASSG method is to describe a global variability and not the local jumps between two points. In addition the distinction between the two possibilities and their impact to a global variability is made by the parameter "travel distance" in Section 6.8. Therefore, the uniqueness of the dynamic progression in the ASSG is not needed.

6.2 Expected Value for ASSGs

With the mathematical fundamental description of the grid and points in the ASSG plot out of the way, it is now possible to formulate new parameters in the ASSG method. The first parameter will be the expected value of an ASSG.

Corollary 6.2.1: Expected value for a single indicator
Let x be an indicator of the ASSG method with a total number of ratings n. Then the expected value of the indicator x is

$$\mathbb{E}(x) \approx \frac{1}{n} \sum_{i=1}^{n} x_i. \tag{6.6}$$

Proof 6.2.1: Expected value for a single indicator
For this proof the corollary of the weak law of large numbers can be used (see eq.
·5.31).To use this corollary, one must show that the indicator x satisfies the weak
law of large numbers.

The set of random variables $(X_i)_{i \in \mathbb{N}}$ are the outcomes of each time point of the
indicator x. This means that the set of random variables $(X_i)_{i \in \{1,...,n\}}$ is the set that
must satisfy the requirements of the weak law of large numbers.

The first premise is that the random variables X_i belong to the same distribution. The rating of each indicator has been interpreted as a random variable and an
unknown probability measure. Therefore, the set of the same random variables X_i
have the same distribution.

The next condition is that $Var(|X_i|) < \infty$. Since the results of each x_i are
between 1 and the highest ranking of the indicator $o \in \mathbb{N}$, the results are limited.
Therefore, the highest possible variance would be $\mathbb{E}((x_i)^2)$ with $\mathbb{E}(x_i) = 0$. The
highest possible value of $\mathbb{E}(x_i)^2$ is when the expected value is equal to the highest
outcome. Therefore, the highest possible variance would be o^2, which is less than
infinity. Thus, $Var(|X_i|) < \infty$ is satisfied.

As an approximation, it is assumed that the random variables are independent.
How good the approximation of the corollary is when the random variables are
not independent is simulated in Section 8.2. Analogous to example 5.6.1 the last
premise of the weak law of large numbers is fulfilled if the set of random variables
is independent (Klenke 2020).

Thus all conditions are fulfilled and the weak law of large numbers can be used
for the corollary of eq. 5.31. This then proves eq. 6.6 which was to be shown.   □

The remaining question is how many evaluations have to be carried out to get a
meaningful result for the approximation of the expected value. Or in other words,
how large is large enough for the weak law of large numbers. To get an answer to this
question, simulations with different distributions for the indicators are examined in
8.2.

Now that the approximation of the expected value of an single indicator in the
ASSG method has been proven, the expected value of an ASSG will be defined by
combining both indicators. The advantage of such a definition is that it is easier to
combine the expected values of both indicators in one parameter.

Definition 6.2.1: Expected value for ASSGs

Let x and y be the indicators of the ASSG method. Then the expected value (EV) for the ASSG plot of these two indicators is

$$EV = \big(\mathbb{E}(x), \mathbb{E}(y)\big)^{T}. \tag{6.7}$$

The second advantage is that now the EV of an ASSG can now also be plotted in the ASSG plot with a red dot, as in Fig. 6.3.

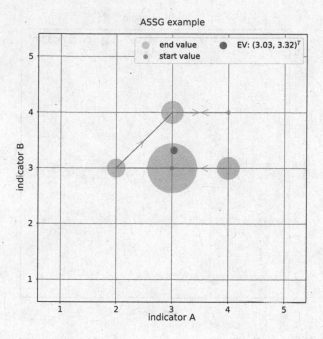

Figure 6.3 ASSG with expected value

Since the EV is the expected value of both indicators, it describes the location where most of the points in the ASSG are located. Therefore, the EV also marks the location of the global attractor, where there is only one attractor. In this particular case, the attractor is located near the point with the longest duration at $(3, 3)^{T}$. It is slightly above it because there are also points above $(3, 3)^{T}$ which are close to the attractor.

This localisation of an attractor is not possible for multiple attractors with only one EV, instead several different EVs would be needed. For example, if there are two attractors, the calculation of the EV has to be done in two divided sets of rankings for one attractor each, which is discussed in the interpretation of the ASSG in Section 7.4.

If the point do not have an attractor as in Fig. 6.3 and are scattered across the whole rating possibilities as in Fig. 6.4, then the expected value cannot be interpreted as an attractor.

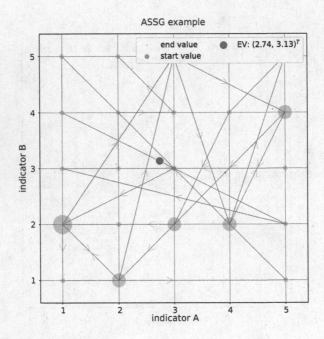

Figure 6.4 ASSG with no attractor

This is only the case if the variability of an ASSG plot is very high. Therefore, it is now necessary to formulate parameters which describe the variability. The aim of the next chapters will therefore be to find such parameters that categorise the variability of all points in the ASSG.

6.3   Standard Deviation for ASSGs

The first two quantities that come to mind as parameters of variability are the variance and the standard deviation. Since they differ only in the scaling of the square root, it makes no real difference which of the two quantities is used. For the case of the ASSG method, the standard deviation is used because a lower value than the variance is helpful for another new parameter in Section 6.5.

Lemma 6.3.1: Standard deviation for a single indicator
Let x be an indicator of the ASSG method. Then the standard deviation $\sigma(x)$ of this indicator is

$$\sigma(x) \approx \sqrt{\frac{1}{n} \sum_{i=1}^{n} \big(\mathbb{E}(x) - x_i\big)^2}. \tag{6.8}$$

Proof 6.3.1: Standard deviation for a single indicator
Let x be an indicator of the ASSG method. With the definition of the standard deviation and the variance (see Section 5.5), the standard deviation can be formulated into combinations of expected values. Corollary 6.6 can then be used to approximate the expected value. This leads to following proof:

$$\sigma = \sqrt{Var(x)}$$
$$= \sqrt{\mathbb{E}\big((\mathbb{E}(x) - x_i)^2\big)}$$
$$\approx \sqrt{\frac{1}{n} \sum_{i=1}^{n} (\mathbb{E}(x) - x_i)^2}.$$

$\square$

6.4   Middle Standard Deviation

Similar to the combined expected value of an ASSG, a standard deviation for an ASSG plot is helpful for an interpretation. The EV also identifies the attractor, because the EV is a vectored parameter. This would not be meaningful for a two dimensional standard deviation, as such a graphical identity as the attractor does not exist in an ASSG plot. Therefore, the global estimate for a variability has to be a scalar. This scalar is a combination of the standard deviations of both indicators. The combination is made by taking an average of the two standard deviations. Given this type of combination, this new parameter is called the mean standard deviation.

Definition 6.4.1: Middle standard deviation
Let x and y be the indicators of the ASSG method. Then the middle standard deviation (MSD) of these indicators is

$$MSD = \frac{\sigma(x) + \sigma(y)}{2}. \tag{6.9}$$

6.5   Deviation Curve

Finally, the mean standard deviation gives an estimate for a global variability. Nevertheless, a graphical object describing global variability is desirable. With such an object, the variability of an ASSG plot can be seen immediately with a single glance at the plot and facilitates interpretation. This object will be called the deviation curve because it is a closed curve in the plot that estimates the deviation of the points in the plot.

Corollary 6.5.1: Deviation curve
Let x and y be the indicators of the ASSG method. Then the deviation curve (DC) of these indicators describes the following closed curve with the form of an ellipse

$$DC(\phi) = \begin{pmatrix} \mathbb{E}(x) + \sigma(x)cos(\phi) \\ \mathbb{E}(y) + \sigma(y)sin(\phi) \end{pmatrix}, \text{ with } \phi \epsilon [0, 2\pi). \tag{6.10}$$

Proof 6.5.2: Deviation curve
The corollary defines a new object, which by itself is only a definition. Only two statements remain to be proven: The object has the shape of an ellipse and it is a closed curve.

For the shape of an ellipse one can use the general form of a parametric description of an ellipse from eq. 5.5:

$$E(\varphi) = \begin{pmatrix} m_x + r_x cos(\varphi) \\ m_y + r_y sin(\varphi) \end{pmatrix}$$

Inserting $\mathbb{E}(x)$ and $\mathbb{E}(y)$ for m_x and m_y as well as $\sigma(x)$ and $\sigma(y)$ for r_x and r_y results in equation 6.10. This shows that the deviation curve has the shape of an ellipse.

To show that the deviation curve is a closed curve, $DC(\phi)$ has to satisfy $DC(0) = DC(2\pi)$ (Bär 2010). This can be shown by a simple calculation:

$$DC(0) = \begin{pmatrix} \mathbb{E}(x) + \sigma(x)cos(0) \\ \mathbb{E}(y) + \sigma(y)sin(0) \end{pmatrix}$$

$$= \begin{pmatrix} \mathbb{E}(x) + \sigma(x) \cdot 1 \\ \mathbb{E}(y) + \sigma(y) \cdot 0 \end{pmatrix}$$

$$= \begin{pmatrix} \mathbb{E}(x) + \sigma(x)cos(2\pi) \\ \mathbb{E}(y) + \sigma(y)sin(2\pi) \end{pmatrix} = DC(2\pi)$$

$\square$

With the additional parameters and the deviation curve, the new ASSG plot can be seen in Fig. 6.5–6.8.

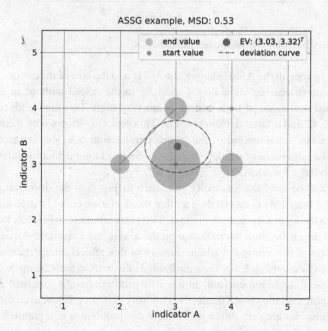

Figure 6.5 ASSG with DC and low variability

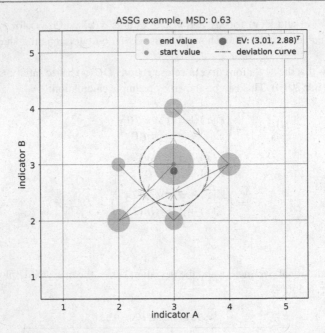

Figure 6.6 ASSG with DC and higher variability

As can be seen in the ASSGs above, the MSD and the size of the deviation curve increase with increasing variability. Especially in the ASSG without an attractor, the standard deviations of both indicators are very high. This also leads to a high MSD of 1.42 and a large deviation curve. This plot also shows why the standard deviation was chosen instead of the variance in Section 6.3. The variance would have led to a larger size of the deviation curve. Such a larger deviation curve would then be too large for ASSG plots like Fig. 6.7.

It can also be seen that in ASSG plots such as Fig. 6.8, the deviation curve is a horizontal line. This is due to the fact that the deviation curve is stretched in the direction of the higher standard deviations. This is also the case in Fig. 6.6. In Fig. 6.8 this is so extreme because the indicator on the x-axis has a standard deviation of 0 since it stays at one rating the whole time. How this affects interpretation will be examined in Section 7.1.2. For the time being, a distinction must be made between this case and a deviation curve as in 6.6. The problem here is that their MSD is the same, although the curves themselves are very different. Equal even though the curves themselves are very different. To solve this problem, a new parameter has to be introduced: the standard deviation difference.

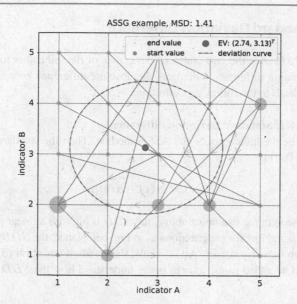

Figure 6.7 ASSG with DC and no attractor

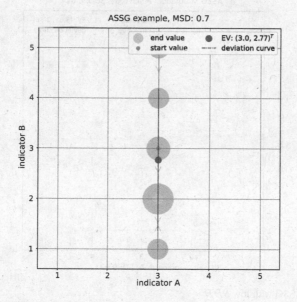

Figure 6.8 ASSG with DC and linear dynamic

6.6   Standard Deviation Difference

While the MSD adds the two standard deviations of the indicators together, the standard deviation difference calculates the absolute difference between the two values. This leads to the following definition:

Definition 6.6.1: Standard deviation difference
Let x and y be the indicators of the ASSG method. Then the standard deviation difference (SDD) is

$$SDD = |\sigma(x) - \sigma(y)|. \qquad (6.11)$$

As it can be seen in the examples above, the SDD is high for a linear progression as in Fig. 6.10 and low for progressions as in Fig. 6.9. Hence, the SDD can distinguish between these two cases. Additionally, the standard deviation (SD) of each indicator will be added to the axis to better understand how the SDD and MSD were calculated.

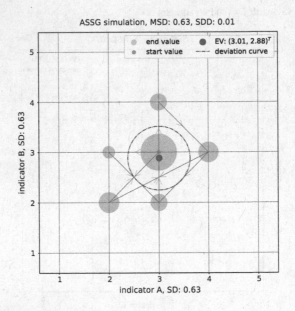

Figure 6.9 ASSG with low SDD

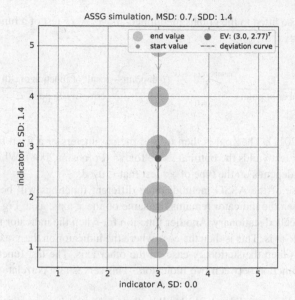

Figure 6.10 ASSG with high SDD

6.7   Fits and Chi-square Tests for ASSGs

As in Fig. 6.10, special types of progressions can occur in an ASSG plot. In the ASSG method, a progression in which one indicator remains the same while the other indicator increases and decreases over time is to be given a graphical object. In addition, a course where both indicators increase or decrease each other shall get a distinguishable parameter. This will be achieved through fits and so called χ^2 tests.

These fits are to integrate three different types of functions into the ASSG plot. These are a linear function, a constant function, and a vertical line. If the points in the ASSG plot match these graphs, then this is a specific type of relationship in the ASSG method, which is discussed in Section 7.1.1 and Section 7.1.2. Whether the points match match these functions is tested with the χ^2 test.

The chi^2 test is a specific type of statistical test to determine how good an estimate of a parameter is (Fornasini 2008). However, there are many different types of χ^2 tests for different types of estimates and estimated parameters (eg. Cochran 1952, Fornasini 2008, Berendsen 2011) In the case of the ASSG method, the parameters tested are those fitted to the graphs mentioned above. A χ^2 test of a function could look like this:

$$\chi^2 = \frac{1}{\text{number of points}} \sum_{i=1}^{\text{number of points}} \frac{(i\text{-th point} - \text{result of function of } i\text{-th point})^2}{\text{error of } i\text{-th point}} .$$

(6.12)

(Berendsen 2011). The goal is then to find the parameters of a fixed function for which the χ^2 test yields the optimal value for χ^2 (Fornasini 2008). Which value is optimal also depends on the type of χ^2 test that is used.

In the case of the ASSG method, three different functions will be used. One would be when the indicator remains the same on the x-axis, as in Fig. 6.10. This case will be called stationary. Another function for when the indicator remains the same on the x-axis. This is then the case where the indicator on the y-axis stays the same. This is then the stationary case on the other axis. The last function is then for a linear function between two indicators. This serves as a correlation test as in Section 3.5.

The first test that will be defined will be the one for the stationary case when the indicator on the x-axis remains the same.

Definition 6.7.1: χ^2 test for stationary fits with fixed x value
Let c be a constant for the stationary χ_x^2 test with fixed x value. Also let x and y be the indicators of the ASSG method with the total number of ratings n. Then the result of the χ_x^2 test is

$$\chi_x^2 = \frac{1}{n} \sum_{i=1}^{n} \left(\frac{x_i - c}{x_i - \mathbb{E}(x)} \right)^2 .$$

(6.13)

This definition is motivated by inserting the corresponding values into eq. 6.12. The number of points for the χ^2 test are the number of ratings of the indicator x, denoted by n. The i-th point is then the i-th rating of indicator on the x-axis x_i. The tested function f is $f(y) = c$ and is the same for every y_i. Therefore, it can be subtracted form x_i by simply using the constant c.

Lastly the error of the i-th point has to be estimated. The i-th rating may have an error. However, the error itself is dependent on the rater of the indicator. Therefore, the error of the rater will be estimated by the variance of the i-th rating of x. The rating on the x-axis is chosen here, because the errors would be the distances to the fixed value of the fitted graph. The graph approximates a situation where each rating of the indicator x stays the same. An error would occur when a rating does not stay the same. One estimate of the deviation of the indicator is the variance. This cannot be estimated from the variance of the rating on the y-axis. The variance of the y-axis cannot describe the deviation from a fixed rating on the x-axis because there is no direct relationship. Therefore, only the variance of the x-axis can estimate the error of the i-th point.

This variance can be calculated as follows:

$$Var(x_i) = \mathbb{E}\big((x_i - \mathbb{E}(x))^2\big) \approx \frac{1}{1} \sum_{j=i}^{i} (x_i - \mathbb{E}(x))^2 = (x_i - \mathbb{E}(x))^2$$

Inserting these values in eq. 6.12 results in eq. 6.13:

$$\chi_x^2 = \frac{1}{\text{number of points}} \sum_{i=1}^{\text{number of points}} \frac{\big(i\text{-th point} - \text{result of function of } i\text{-th point}\big)^2}{\text{error of } i\text{-th point}}$$

$$= \frac{1}{n} \sum_{i=1}^{n} \frac{(x_i - f(x_i))^2}{Var(x_i)} = \frac{1}{n} \sum_{i=1}^{n} \frac{(x_i - c)^2}{(x_i - \mathbb{E}(x))^2} = \frac{1}{n} \sum_{i=1}^{n} \Big(\frac{x_i - c}{x_i - \mathbb{E}(x)}\Big)^2$$

In order to get a further understanding of this χ_x^2 test a simple example will be made.

Example 6.7.1: χ^2 test for stationary fits with fixed x value

Let x be an indicator with fixed ratings at $x_i = 2$ for all ratings. In addition, let y be an indicator with a random rating between 1 and 5. Then a χ_x^2 test should lead to an optimal χ_x^2 value for the function $f(y) = 2$. With $c = 2$ and each $x_i = 2$ in the χ_x^2 test of eq. 6.13 the sum results in the summation of zeros. Therefore, χ_x^2 should be zero for this set of ratings. Looking at the plotted χ_x^2 test of Fig. 6.11 it also yields $f(y) = 2$ and $\chi_x^2 = 0$. Here, a program was used to test various parameters c for their result of a χ^2 test. Then the function is plotted for the optimal value of the test, in this case 0.

For an x-indicator with a higher variability the result of χ_x^2 increases as in Fig. 6.12. This is because then the summation over the differences between c and x_i are not always zero.

As the example above shows, the χ^2 value calculates how big the difference is between the optimal curve and the tested points. This difference is then squared and normalised by the estimated error of each point. Such a difference is then calculated and summed for each point in the test. The results are then normalised again by

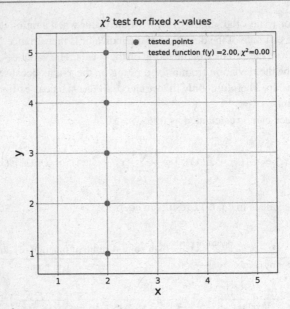

Figure 6.11 χ_x^2 test with perfect alignment

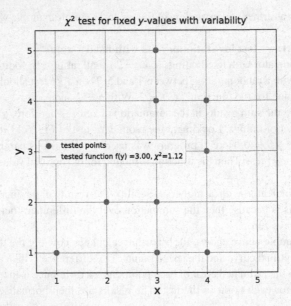

Figure 6.12 χ_x^2 test with variability

the number of points in the summation. Therefore, the χ^2 test is an estimate of the general quadratic difference between the optimal function and the points tested.

The optimal result for χ_x^2 in the defined test is zero. This result can only be obtained when all points are on the optimal function. How large χ_x^2 may be for a function to still be considered optimal is unknown. How big χ_x^2 is allowed to be for a function to still be considered optimal remains unknown. Therefore, the values of the χ^2 tests have to be simulated in order to make an estimate of the critical value of these χ^2 is. Such an simulation will be carried out in Section 8.3.

The same kind of χ^2 test for a fixed y value can be formulated analogously to the χ_x^2 test. Here, the ratings y_i will take the role of x_i. Their variance is also $(y_i - \mathbb{E}(y))^2$. This results in the following χ_y^2 test:

Definition 6.7.2: χ^2 test for stationary fits with fixed y value
Let c be a constant for the stationary χ_y^2 test with fixed y value. Also let x and y be the indicators of the ASSG method with n ratings. Then the result of the χ_y^2 test is

$$\chi_y^2 = \frac{1}{n} \sum_{i=1}^{n} \left(\frac{y_i - c}{y_i - \mathbb{E}(y)} \right)^2. \tag{6.14}$$

A similar example as before for the χ_x^2 test can be done too.

Example 6.7.2: χ^2 test for stationary fits with a fixed y value
This time, let y_i be fixed at the rating 2 and the indicator x jumps between all ratings, thus χ_y^2 should be zero again since the summation is done over zeros. Again, for any variability, χ_y^2 is near zero (Fig. 6.14).

These two χ^2 test can also be integrated into the ASSG method. For the ASSG plot, both tests are performed for each diagram, even if the orientation of the points does not fit for the diagram. This is done to get more information about the ASSG plot without having to examine the orientation of the points.

In Fig. 6.16, where the points are almost perfectly aligned for a stationary fit with fixed x value, the result of the χ_x^2 test is almost zero. For the χ_y^2 test, the result is very high. This is intended, because a fit for a fixed y value where the x value is fixed instead should not lead to an optimal χ_y^2 result.

The limit case would be for points that have perfect alignment, as in Fig. 6.15, where the χ_x^2 test results in zero, just as in the example in Fig. 6.13. However, the χ_y^2 value is infinity (short inf). The reason for this is the standard deviation in which each value in the sum in divided by. If there are no variabilities for the ratings y_i then the standard deviation of every point is zero. Therefore, the terms in the sum

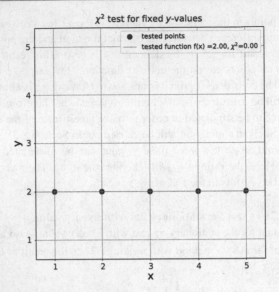

Figure 6.13 χ_y^2 test with perfect alignment

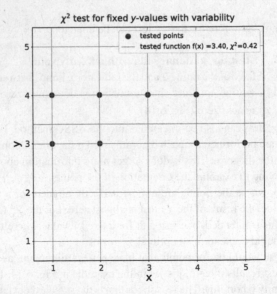

Figure 6.14 χ_y^2 test with variability

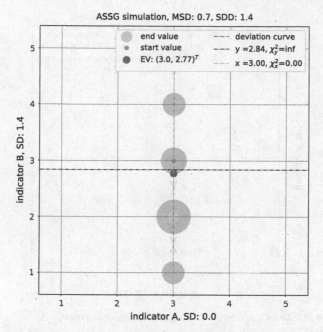

Figure 6.15 ASSG with χ_y^2 and χ_x^2 tests and perfect alignment in its points

are divided by zero, which is not defined. Therefore, the division is done by a term with limit zero. This leads to a summation over infinity, which results in infinity. This is a drawback of this type of χ^2 test, because the values for χ^2 are not limited. For an interpretation of a single ASSG plot it is not a problem if only one of the three χ^2 tests results in infinity. For the interpretation, only low values of the chi^2 are important.

The problem lies in a simulation for the critical values of the χ^2 test, because a mean value across the entire simulated data will result in infinity if only one of the simulated χ^2 is equal to infinity. Therefore, this problem has to be addressed in Section 8.3.

The last fitted function in the ASSG method is a linear function. The χ^2 test for a linear function is similar to the already defined test. In this case, the result of the tested function f has to be calculated for every y_i independently.

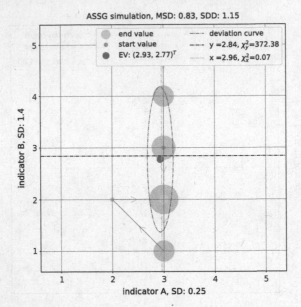

Figure 6.16 ASSG with χ_y^2 and χ_x^2 tests and variability in its points

Definition 6.7.3: χ^2 **test for linear fits**
Let f be a linear function for the χ^2 test. Also let x and y be the indicators of the ASSG method with a total number of n ratings. Then the result of the χ^2 test is

$$\chi_{lin}^2 = \frac{1}{n} \sum_{i=1}^{n} \left(\frac{y_i - f(x_i)}{x_i - \mathbb{E}(x)} \right)^2. \tag{6.15}$$

Example 6.7.3: χ^2 **test for linear fits**
As in the last two examples, the χ_{lin}^2 test yields the optimal value $\chi_{lin}^2 = 0$ for a perfect alignment. With more distortions the χ_{lin}^2 value increases. The difference between the χ_{lin}^2 test, the χ_y^2, and the χ_x^2 test is that here two parameters for the function $f(x) = mx + b$ have to be tested. These are the slope of the linear function and the intercept of the y-axis (Fig. 6.17, 6.18).

With the additional χ_{lin}^2 tests an ASSG with all the χ^2 tests looks like Fig. 6.19 and Fig. 6.20. Similar to the example of the χ_{lin}^2, is the χ_{lin}^2 value low for linear alignments like in Fig. 6.20. In contrast, the χ_{lin}^2 value increases for an ASSGs with no linear alignment as in Fig. 6.19.

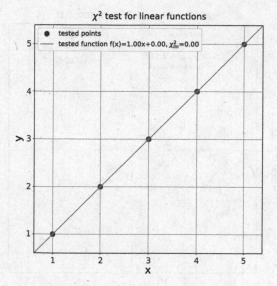

Figure 6.17 χ^2 test for linear functions with perfect alignment

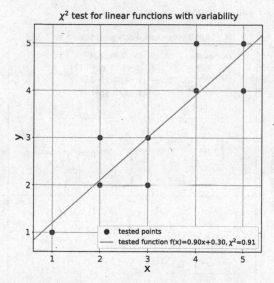

Figure 6.18 χ^2 test for linear functions with variability

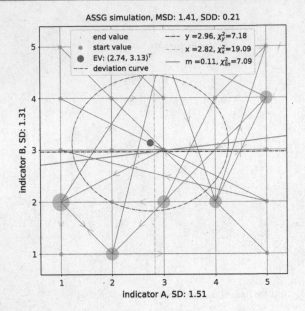

Figure 6.19 ASSG with all χ^2 tests

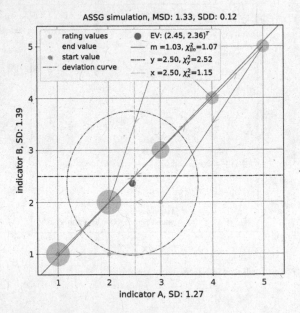

Figure 6.20 ASSG with linear alignment

6.8   Travel Distance

The next problem that arises is that the variability cannot be distinguished between ASSGs that frequently jump between the same points or stay there longer. The parameter MSD alone remains the same in both cases, as in Fig. 6.21 and Fig. 6.22.

This is the same problem that was already discussed in Section 6.1 where the ASSG plot cannot distinguish among the many jumps between $(3, 3)^T$ and $(4, 3)^T$ and the long continuous duration in $(3, 3)^T$ of Fig. 6.23.

To solve this problem, a new parameter is needed that estimates the average distance between a point and the next point in time. If there are many jumps between points, this parameter should result in a higher average distance than in an ASSG where the points stay a longer duration at the same spot. Therefore, such a parameter describes the average distance a point travels between an evaluation time interval. Therefore, this parameter is called travel distance.

Definition 6.8.1: Travel distance
Let x and y be the indicators of the ASSG with a total number of n ratings. Then the travel distance (td) of the ASSG method is

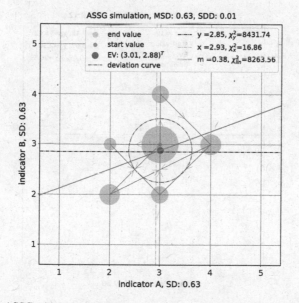

Figure 6.21 ASSG with low variability

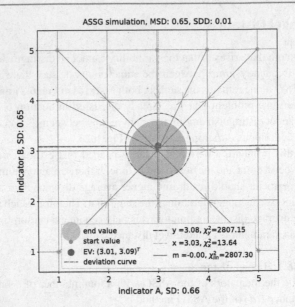

Figure 6.22 ASSG with high variability

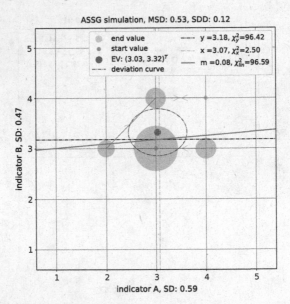

Figure 6.23 ASSG of Section 6.1 with all new parameters so far

$$td = \frac{1}{n-1} \sum_{i=1}^{n-1} |(x_i, y_i)^T - (x_{i+1}, y_{i+1})^T|. \tag{6.16}$$

The travel distance is a mean value for all distances between two successive points in time. The distance between the i-th point and the $i + 1$-th point is $|(x_i, y_i)^T -$ $(x_{i+1}, y_{i+1})^T|$. Since this distance can only be calculated between two points, the possible number of distances that can be calculated with n ratings is $n-1$. Therefore, the summation for the mean value is only across $n - 1$ terms. This then leads to equation 6.16 above.

Now the ASSG method can distinguish between the variability of Fig. 6.25 and Fig. 6.24.

Here the average distance of the individual points in Fig. 6.24 is higher than in Fig. 6.25. This can be seen in the travel distance of both ASSG, with the right one being 0.24 longer. Therefore, the points in Fig. 6.24 travel on average 0.24 longer than in the left plot.

In addition, the travel distance of Fig. 6.26 can estimate whether there are many jumps between $(3, 3)^T$ and $(4, 3)^T$ or not. The difference between the travel

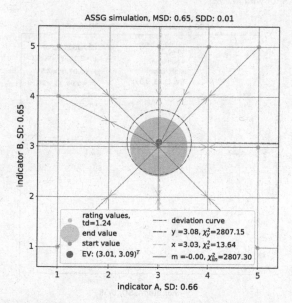

Figure 6.24 ASSG with high td

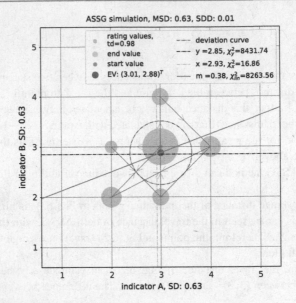

Figure 6.25 ASSG with low *td*

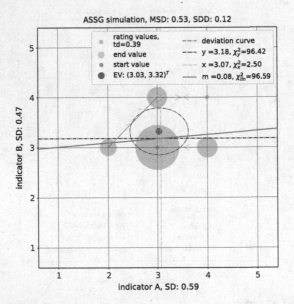

Figure 6.26 ASSG of Section 6.1 with travel distance

distances of Fig. 6.26 and Fig. 6.25 is 0.53, which is even larger than that between the last two. Although the total number of rating combinations in both ASSGs differs only by one. Therefore, there cannot be as many jumps in Fig. 6.26 as in Fig. 6.25. Since the difference of tds is so large, this suggests that there cannot be many jumps between $(3, 3)^T$ and $(4, 3)^T$.

6.9   Inner Point Density

Another desirable distinction in the ASSG method is how many points are located within the deviation curve. This cannot be distinguished with the existing parameters. While the area of the deviation curve becomes larger the more points have a larger distance to the tractor, this does not mean that the number of points within the deviation curve remains constant. This can be seen in the difference between Fig. 6.26 and Fig. 6.27.

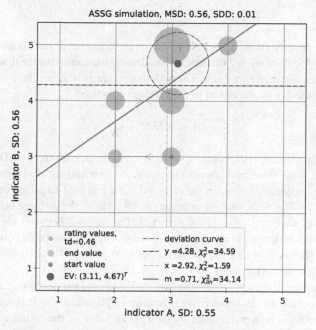

Figure 6.27 ASSG with few points inside the deviation curve

In this case, the MSD remains the same and therefore the area of the deviation curve is the same in both ASSGs. It can be argued that the path distances of the two ASSGs are different here and that this could be different in both cases. However, as argued in the last chapter, if the points between $(3, 3)^T$ and $(4, 3)^T$ jump more frequently in Fig. 6.26 then the travel distance could be a high as in Fig. 6.29. Thus, the travel distance between these two cases cannot be different either. Therefore, a new parameter is needed that describes the number of points within the deviation curve. For this purpose, the number of points within the curve is counted and divided by the total number of points. This gives an estimate of the density of points within the deviation curve. Therefore, this parameter is called "inner point density".

Definition 6.9.1: Inner point density
Let x and y be the indicators of the ASSG with a total number of ratings n. Then the inner point density (ipd) of the ASSG method is

$$ipd = \frac{1}{n} \sum_{i=1}^{n} \mathbb{1}_{\left\{ i \in \{1,...,n\} \mid \overline{|EV-DC|} > |EV-(x_i, y_i)^T| \right\}} (i). \qquad (6.17)$$

Similar to the definition of the size of the circle diameter at one location of the grid in the ASSG of eq. 6.2 the inner point density uses an indicator function that distinguishes between two cases. The difference here is whether the i-th point in the ASSG satisfies the statement in the indicator function:

$$\overline{|EV - DC|} > |EV - (x_i, y_i)^T|.$$

This will approximate if the point $(x_i, y_i)^T$ is inside the deviation curve. This would be the case if the distance between the expected value and the deviation curve is larger than the distance between the expected value and the i-th point. Such a case can be seen in Fig. 6.28 with the point at $(3, 3)^T$. There, the distance between the point and the expected value is coloured red. The distance between the expected value and the deviation curve is coloured orange. Since the distance between DC and EV (0.80) is larger than the distance between point and EV (0.36) the point lies within the deviation curve.

However, the distance between the expected value and the deviation curve is not constant, because the deviation curve is an ellipse. Therefore, the distance between these two objects is approximated by their mean value of the distance. The difference between the mean and the actual distance can be seen between the blue deviation curve and the orange circle in Fig. 6.28. This approximation struggles in

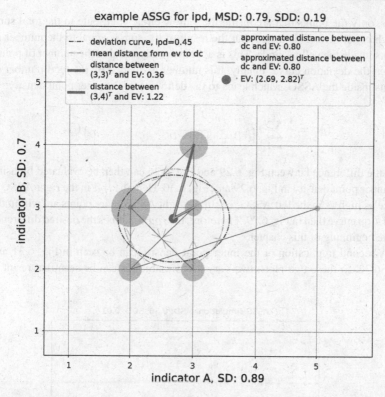

Figure 6.28 Construction of the distances in the definition of the inner point density

long stretched deviation curves. This is the case because a circle is a good approximation for an ellipse only if the ellipse is similarly elongated in both directions. This has to be kept in mind for an interpretation of the inner point density.

In contrast to the point at $(3, 3)^T$, the distance between EV and DC smaller than the distance between $(3, 4)^T$ and the EV. This means that this point does not satisfy the statement of the indicator function. Hence the indicator function results in zero for this term in the sum. This procedure is then carried out with all points within the ASSG. This leads to the following indicator function within the definition of the inner point density:

$$\mathbb{1}_{\left\{i \in \{1,...,n\}| \overline{|EV-DC|} > |EV-(x_i,y_i)^T|\right\}}(i).$$

Thus, only the points that are inside the yellow circle contribute to the total sum, which is not zero. This means that the result of the sum is equal to the number of points inside the yellow circle. This is approximately equal to the number of points within the deviation curve. Finally, this number is divided by the total number of points inside the ASSG, which leads to the definition of the inner point density:

$$ipd = \frac{1}{n} \sum_{i=1}^{n} \mathbb{1}_{\left\{ i \in \{1,...,n\}| \overline{|EV-DC|} > |EV-(x_i,y_i)^T| \right\}} (i).$$

The the difference between Fig. 6.29 and Fig. 6.26 can then be evaluated by using the inner point density in Fig. 6.29 and Fig. 6.30. Here the ipd in the right ASSG is twice as high as in the left ASSG, i.e. in the right ASSG more points are within the deviation curve than in Fig. 6.29. Therefore, the ipd describes the desired difference at the beginning of this chapter.

A second implication of the inner point density can be seen in Fig. 6.31 and Fig. 6.32. In these two plots, two very low variabilities can be seen that result in

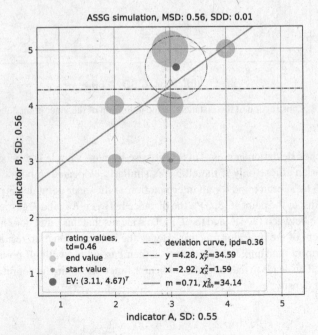

Figure 6.29 ASSG with low ipd

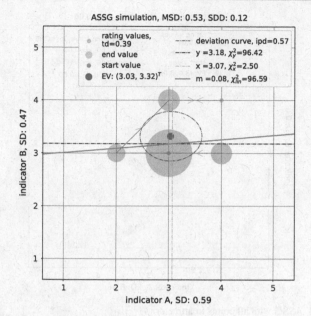

Figure 6.30 ASSG with high *ipd*

almost all points being within the same location. How many points are inside this location in contrast to all existing points is shown in the *ipd*.

In Fig. 6.31 the ipd is 0.96. With a total number of $n = 25$ ratings this means that

$$n \cdot ipd = 25 \cdot 0.96 = 24$$

points are inside the deviation curve and

$$n - n \cdot ipd = 25 - 24 = 1$$

points are not inside the deviation curve. The deviation curve itself is a very small ellipse, which can barely be seen in the plot because the MSD is too low (0.11). In contrast to this ASSG there are $25 \cdot 0.84 = 21$ points inside the deviation curve of Fig. 6.32. Resulting in a lower *ipd*. Therefore, the *ipd* also helps in counting the number of points within a location if the variability is low enough. This will be helpful for the interpretation of "pseudo relations" in Section 7.1.3, where the

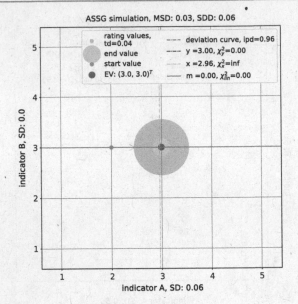

Figure 6.31 ASSG with all points in attractor

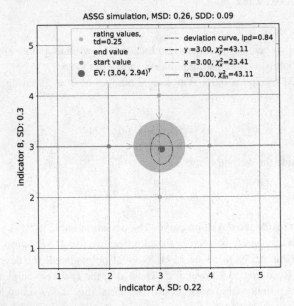

Figure 6.32 ASSG with most points in attractor

variability of the indicator could also be caused by a poorly defined ranking system or coincidence.

6.10   Travel Tendency

The last parameter in the ASSG method is intended to describe whether the ratings of the indicators increase or decrease over time. This could be the case, for example, if a teacher is still nervous in the first part of a lesson and gains more confidence in the later part of the lesson. Such a lesson could result in the scores of an indicator improving over time. Of course, this depends on the type of indicator used, however an indicator that depends on the confidence of a teacher could have such a dynamic course. Therefore, a parameter describing the tendency of the directions in which the points of an ASSG travel would be helpful for such an situation. This parameter will be called travel tendency and is defined as follows:

Definition 6.10.1: Travel tendency
Let x and y be the indicators of the ASSG. Then the travel tendency (tt) of the ASSG method is

$$tt = \frac{1}{2(n-1)} \sum_{i \in \mathcal{I}} \left(\mathbb{1}_{\{i \in \mathcal{I} | x_{i+1} - x_i > 0\}}(i) + \mathbb{1}_{\{i \in \mathcal{I} | y_{i+1} - y_i > 0\}}(i) \right.$$

$$\left. - \mathbb{1}_{\{i \in \mathcal{I} | x_{i+1} - x_i < 0\}}(i) - \mathbb{1}_{\{i \in \mathcal{I} | y_{i+1} - y_i < 0\}}(i) \right) \qquad (6.18)$$

with $\mathcal{I} = \{1, ..., n-1\} \subset \mathbb{N}$.

This definition is very similar to the definition of the inner point density mentioned in the last chapter. However, the difference is that there are six cases for the travel tendency:

1. the x component of the ith point travels to a higher value resulting in

$$x_{i+1} - x_i > 0,$$

2. the y component of the ith point travels to a higher value resulting in

$$y_{i+1} - y_i > 0,$$

3. the x component of the ith point travels to a lower value resulting in

$$x_{i+1} - x_i < 0,$$

4. the y component of the ith point travels to a higher value resulting in

$$y_{i+1} - y_i < 0,$$

5. the x component of the ith point remains the same value resulting in

$$x_{i+1} - x_i = 0,$$

6. the y component of the ith point remains the same value resulting in

$$y_{i+1} - y_i = 0.$$

The first case would result in an one in the first indicator and zero in the other three indicators. The second case would result in an one in the second indicator. The third indicator is one in the third case resulting in a -1 for the sum. The same result applies to the fourth case and the fourth indicator. The last two cases result in a zero for each indicator in the sum. Therefore, the first two cases increase the sum for the travelers, while the third and fourth cases decrease the sum by one. The last two cases do not affect the sum because each indicator remains zero. Thus, the sum increases when the score of an indicator increases, and it decreases when the indicator decreases. Otherwise, the sum does not change. This process will be done for all the points in the ASSG and the summed results are then normalised. The highest outcome would be if all indicator functions with a positive sign sum up to 1. The result would then be double the number of tested combinations of points $n - 1$. The lowest possible result would be if all indicators with a negative sign result in 1. Then the result would be $-2(n - 1)$. Therefore, the value norm of this parameter is $2(n - 1)$. This results in the above definition of the traveling tendency.

Positive values for the travel tendency indicate that the rating progression tends toward increasing values of the ratings. Negative values, on the other hand, indicate a decrease in the values of both indicators. A value close to zero indicates that the dynamic trend has no effect in any direction. If the result of tt is equal to 1, then all ratings progress in a positive direction. The same is true for $tt = -1$ when all ratings progress in a negative direction.

An ASSG with a high travel tendency of -0.10 can be seen in Fig. 6.33. This means that the points in the ASSG decrease on average with a rating of 10%.

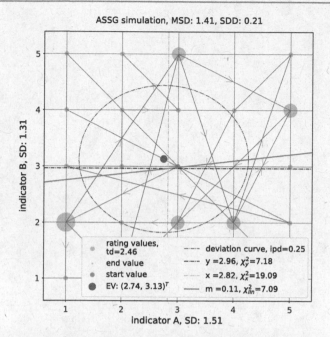

Figure 6.33 ASSG with all new parameters and high travel tendency

After defining the last new parameters in the ASSG method, their interpretation potential for indicator-based measurements can be explained in the next chapter.

Interpretation of the ASSG Plots

7

The first objective of this chapter is to categorize the different types of possible relationships for an ASSG. This will be a first attempt towards answering the research question "How can different types of SSG plots be categorized?". In order to do this, Section 7.1 describes the different types of relationships. Then, in Section 7.2, a flow chart is created for an initial categorisation. The critical values for the parameters in this flow chart will be simulated later in Section 8.3. Furthermore, another application for disturbances is explained in Section 7.4. Finally, the requirements for indicators in a student-teacher system are discussed in Section 7.5.

7.1   Relation Types

The following chapters first describe the different types of relationships in an ASSG. The first three types are special cases such as the linear relationship in Section 7.1.2. The fourth and fifth types are the relationship types where the indicators are influenced by an attractor. The last relationship type is the one where the ASSG method cannot find a specific type of relationship.

7.1.1   Stationary Relations

The first type of relationship in the ASSG method is called a stationary relation. In this case, the points in the ASSG arrange linearly, as shown in Fig. 7.1. As can be seen in the ASSG plot above, the indicator on the x-axis remains the same while the indicator on the x-axis changes. This means that the rating in the indicator B

N. Litzenberger, *Introduction of Advanced State Space Grids and Their Application to the Analysis of Physics Teaching*, BestMasters,
https://doi.org/10.1007/978-3-658-42732-0_7

remains stationary in the same rating throughout the time of all ratings. It could therefore be the case that the variability of the indicator A does not have an impact on the rating of indicator B. However, this would mean that it is a coincidence that indicator B does not change. Another possibility is that the variability of indicator A causes indicator B to stay the same. It is not possible to determine which of these two possibilities is true with a single ASSG. To further analyse this type of relationship, different lessons need to be assessed. This would reduce the probability of the first possibility, as it is unlikely that the score for different lessons would always remain the same if the effect of this type of relationship is random. The only possibility left would be that indicator B remains the same because indicator A changes. Despite the variability of A, B remains constant.

It is also possible that the ratings of indicator B are poorly defined, so that they would never show any variability. This could be the case, for example, if other ratings are highly unlikely to happen. This should be considered when formulating an indicator for this method. Such requirements for indicators in the ASSG method will be covered in detail in Section 7.5.

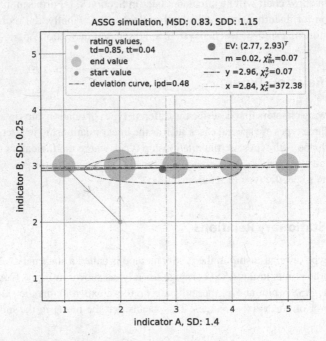

Figure 7.1 ASSG with stationary relation with fixed y-axis

With the parameters of the ASSG method, this kind of relation can be found by the χ_x^2 or χ_y^2 test. This is the case because if an indicator rating remains the same, the points of the ASSG method remain on the horizontal or vertical line. This type of alignment is exactly the type for which the χ_x^2 and χ_y^2 were defined. Therefore, the χ_x^2 value is close to zero when the indicator hardly changes on the x-axis. Similarly, the χ_y^2 value is close to zero if the indicator hardly changes on the x-axis. If the indicator does not change in one of the axes, the χ_x^2 or χ_y^2 value is zero, as in Fig. 7.1.

An example of a stationary relation where the χ_x^2 yields close to zero can be seen in Fig. 7.2. Here the roles of the indicators A and B are reversed. Now indicator A remains the same while indicator B changes.

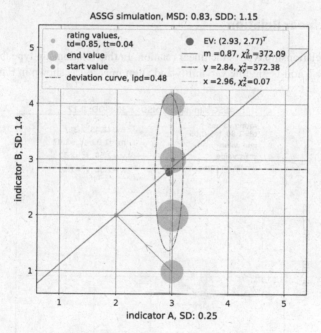

Figure 7.2 ASSG with stationary relation with fixed x-axis

In contrast to the example in Fig. 7.1, the χ_{lin}^2 test does not give the same value as the χ_x^2 test. This is due to the fact that the χ_{lin}^2 test is optimised for the linear relation from Section 7.1.2, where the parameter m for the test function $f(x) = mx + b$ has to satisfy $m \approx \infty$ to result in a χ_{lin}^2 value of almost zero. The testable parameters for m are limited and cannot increase to such high values. Nor is this necessary, because the χ_x^2 test is defined precisely for this case of Fig. 7.2. Therefore, the χ_{lin}^2 test does not have to cover this case.

This stationary type can be further differed with the parameter tt. If the absolute value of tt is significantly greater than zero, then the dynamic course of the ASSG plot has a tendency towards one direction of the ratings. Since an indicator is always the same, this progression can only apply to the changing indicator. Therefore, this indicator has to increase (if $tt > 0$) or decrease (if $tt < 0$) over time. If this type of relationship does not occur by chance or due to a poorly defined indicator, then this implies that the changing indicator is increasing/decreasing over time because the other indicator remains the same.

7.1.2   Linear Relation

The next relationship type is the linear relation. In this relationship type, both indicators have a non-negligible variability and align linearly, as shown in Fig. 7.3.

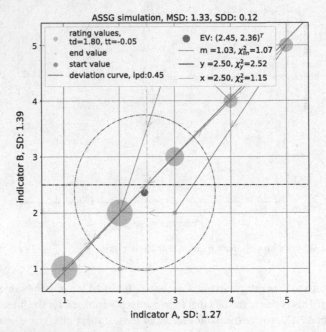

Figure 7.3 ASSG with linear relation

In this case, if one indicator increases, the other also increases. The same applies if one indicator decreases, then the other also decreases. This leads to a linear alignment as in Fig. 7.3.

As with the stationary relation, a χ^2 test can be found for this type of relationship. In this case it is the χ^2_{lin} test. If the result of the χ^2_{lin} test is close to zero, the indicators are linearly related, as in the example above. In this case, the value for the test function is greater than zero. This means that if the property of indicator A improves, then the property of indicator B also improves. If the property of indicator A deteriorates, the property of indicator also deteriorates. If the value for m is less than zero, the opposite is true. That is, if the property of indicator A improves, the property of indicator B also deteriorates. The same applies in the opposite direction.

This type of relationship is specific to $m > 0$ and a high absolute value for tt. If $tt < 0$, both indicators decrease over time, and if $tt > 0$, both indicators increase over time. This type of relationship cannot be found for $m < 0$. This is because for every indicator function with a positive sign that yields a one, there is an indicator function with a negative sign that yields a one in the definition of tt (see eq. 7.6). Therefore, the result for tt remains close to zero.

The question that remains is what high and low the results are for the travel tendency. To answer this question, the values of this parameter are simulated in Section 8.3.

7.1.3 Pseudo Relation

The opposite of the linear relation, where the variability of one indicator's rating can influence the variability of the other indicator's rating, is the pseudo relation. This is the case when both indicator ratings have such a variability that they remain in one place. Such a type of ASSG can be seen in Fig. 7.4.

Here the ipd is close to one since almost all points are in the same place. There is only one point in $(2, 3)^T$ and one point in $(3, 4)^T$ that are not within the deviation curve. This means that both indicators only change for one time interval.

This could be a coincidence in the sense of a stationary relation and needs to be investigated with further ASSGs. Is could also be the case that both indicators are defined in a way that makes it impossible to display a dynamic progression (see Section 7.1.1 and Section 7.5). This could mean that they are not linked. It is also possible that both indicators are so strongly related that the localisation of one of the indicators at one point forces the other indicator to remain at its location as well. However, their relationship type prohibits dynamic development. This would mean that they are strongly interrelated.

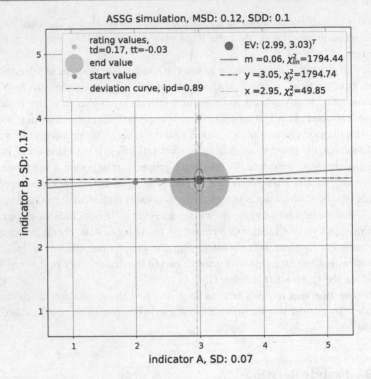

Figure 7.4 ASSG with a pseudo relation

This is the "worst case scenario" for an ASSG, because only further ASSGs of different lessons can solve this problem. The information gain here is minimal, because anything between a strong relation type or no relationship at all is possible.

7.1.4 Strong Relation

In the case of the strong relation, the ratings are more spread out than in a pseudo relation. The indicators have a low variability and a strong attractor that leads to a location in the ASSG that "attracts" the points around it. A strong attractor means that a high percentage of the points are within the location of an attractor. An example of a strong relation can be seen in Fig. 7.5.

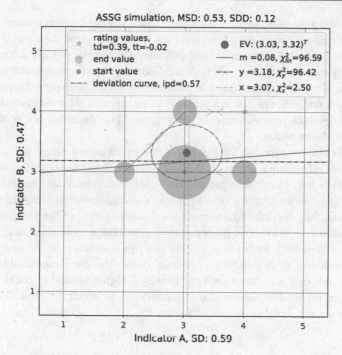

Figure 7.5 ASSG with a strong relation

In this case the variability is higher than in a pseudo relation. Therefore, some of the ratings in the ASSG plot are not located in an attractor. In Fig. 7.5 the attractor is at the expected value of $(3.07, 3.29)^T$. The deviation curve wraps around this attractor and has a significantly lower ipd than the peseudo relation of Fig. 7.4. This is because the variability of the ASSG is so high there that more indicators are outside an attractor. Consequently, the MSD increases. Therefore, this case leads to an ipd lower than than a pseudo relation and to a MSD higher than a pseudo relation.

For the interpretation of the indicators, this means that by one indicator being close to a stable rating in the attractor, the other indicator is also attracted to the attractor. If a rating is outside the attractor, it comes back into the attractor after a short time. An example of this is the jumps between $(3, 3)^T$ and $(4, 3)^T$ in Fig. 7.5. The indicators thus stabilise each other in this system.

How often this stabilisation process is done can be seen in the travel distance. This is because this parameter describes how often such jumps occur between the attractor and a location outside the attractor. Thus, if a stabilisation process occurs in which a rating jumps away from and back into the attractor, the travel distance increases. The SDD on the other hand estimates whether this process occurs more frequently for an indicator or not. Each stabilisation process increases the standard deviation of one indicator, because the rating changes form the expected value of the indicator. If one indicator deviates further from its expected value than the other, the standard deviation increases more in comparison. Thus, the difference in standard deviation also increases.

This is also the best case for a global evaluation through the expected value. In this type of relationship, the standard deviations of both indicators are low because the MSD is low. Thus, the expected value has low deviations and represents the overall rating better than in a case where the standard deviation is higher. On the other hand, they also take into account the dynamic course of the rating through their standard deviation. This is not possible in the case of a pseudo-relation.

Higher variability makes it less likely that this correlation is due to a poorly defined indicator, as they display a dynamic course. This correlation could still be random, although it is rather unlikely. However, higher variability could further reduce this probability. Such a type of relationship, in which the attraction to the attractor is weak, is called a weak relation.

7.1.5 Weak Relation

The weak relation has a higher variability than the strong relation. This means that the attractor is less "strong" than in a strong relation. Therefore, the points in the ASSG plot are still attracted to the attractor. However, there are more points outside the attractor. This leads to a higher MSD and lower ipd as in Fig. 7.6.

The travel distance indicates how often the process is carried out. The standard deviation process describes whether one indicator travels through the process more often than the other. In the case of Fig. 7.6, the process occurs more often than in Fig. 7.5, however this is also due to the fact that the attractor concentrates the points more to the expected value of the ASSG plot in a strong relation. The SDD is the same in both ASSGs, therefore the tendency of one indicator going through this process more often than the other remains the same.

In contrast to the strong relation, the variability of the indicators is higher. This means that the influence that one indicator has on the other leads to a lower concentration of points within an attractor. In addition, the system of the two indicators is less stable because the scores change more frequently. How often they change can be described through their travel tendency. However, this type of relation has

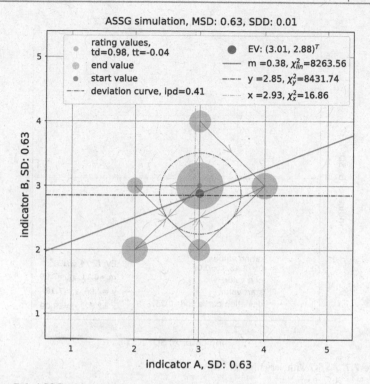

Figure 7.6 ASSG with weak relation

a lower probability of being a random event because the dynamic process around the attractor is more variable. This means that the attraction of the attractor has to happen more often randomly. This is less likely than random alignment in a strong relation. This is an advantage that the weak relation has over a strong relation.

7.1.6 No Relation

The next logically following case of an ASSG plot would be a case with even higher variability. This would lead to an even bigger area of the deviation curve and even fewer points within the curve. This also leads to an even higher MSD. Such an ASSG plot can be seen in Fig. 7.7.

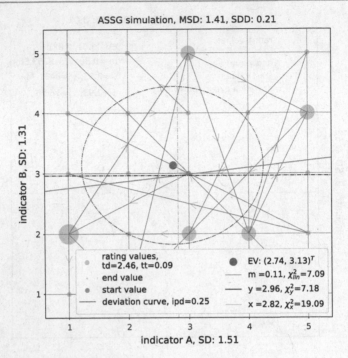

Figure 7.7 ASSG with very high MSD

In this case, the indicator ratings are not attracted to a particular point. Therefore, the expected value is in the middle of the graph and the deviation curve covers a large area around it. Since there is no attractor in the ASSG plot the expected value does not serve a purpose. This also means that there is no stable combination for both indicators in the student-teacher system.

While it could be the case that variability in one indicator also indicates high variability in the other indicator. However, there is no other way to describe the nature of the dependence of the indicators on each other in the ASSG method. The only thing that can be estimated is how high the variability is. The MSD describes the global variability of the system. The travel distance describes how large the distance between two ratings is on average. And the SDD describes whether one indicator has a larger deviation than the other.

It is not possible to estimate more than this with the ASSG method. Therefore, the ASSG can not find a specific relation type. Therefore, such an ASSG type is called a no relation type in the ASSG method. This does not mean, that the indicators can not be related at all. It only means that the ASSG method has no parameters to clearly categorise this relationship type. This is nevertheless valuable information for interpretation. In this case, one piece of information would be that the indicators are not stable during the assessment period. The other is that they do not correspond to any of the predefined relationship types from the previous chapters.

An other example where the ASSG method cannot find the relation type is the case in Fig. 7.8. In this ASSG are two attractors. One is located at $(5, 5)^T$ and the other is at $(1, 1)^T$.

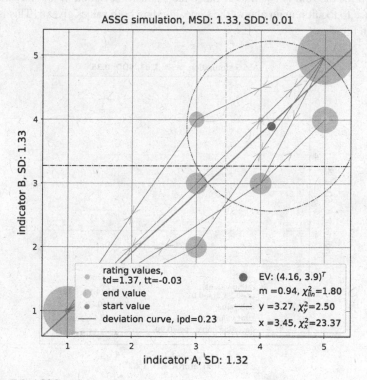

Figure 7.8 ASSG with two attractors

Since there is only one global expected value the EV of an ASSG can not find the correct location of any of the two attractors. Instead, the EV is attracted more to the attractor with more points inside (in this case this would be $(5, 5)^T$). If both attractors are equally concentrated, then the EV would be in the middle of both. The only categorisation that can be done is with the χ^2_{lin} test, which results in $\chi^2_{lin} = 1.80$ in this case. This is because the points between the first and second attractor still align linearly.

Such a categorisation can only be made with two attractors. An ASSG with four attractors as in Fig. 7.9 does not have a linear alignment, which means that the χ^2_{lin} test cannot find a relationship either. However, if the number of points on the left and right sides of a stationary fit with a fixed x is almost the same, a χ^2_x test results in low values. This is due to the fact that the χ^2 test leads to a result when every element of the sum is equal to 1. Since the standard deviation of a point and the distance to the fit remain almost the same, dividing both values gives 1. This is the case in Fig. 7.9.

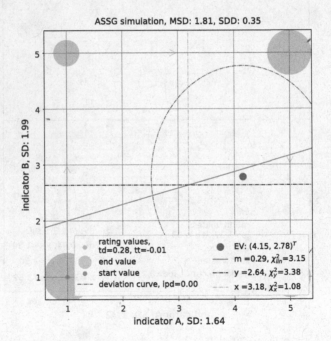

Figure 7.9 ASSG with four attractors

In order to describe this type of relationship correctly, several EVs would be required, one for each attractor. How such an ASSG might look like will be covered in Section 7.4. What applications this might have for interpreting lessons is also described there.

Nevertheless, it is not possible to distinguish between the types of Fig. 7.7 and Fig. 7.9 without first looking at the ASSG plot. Therefore, it is always necessary to examine the ASSG plot when the parameters describe a no relation type in the ASSG method. The advantage of the ASSG method is that the examination of the ASSG plot is only necessary for the no relation type to determine essential differences in the plots. Any other relation types mentioned before can be categorised with the parameters of the ASSG method. How such a categorisation comes about is substantiated for the first time in the next chapter.

7.2   Categorisation for Relation Types

The aim of this chapter is to establish an initial categorisation for the six relationship types of Section 7.1 without having to look at the ASSG plot itself. This has the advantage that not all ASSG plots have to be examined when searching for a certain relation type. In addition, an initial categorisation of all ASSGs makes interpretation easier when one has many indicators. For example, an assessment of six indicators yields $6 + 5 + 4 + 3 + 2 + 1 = 21$ combinations for ASSG plots. An initial categorisation saves more time in interpretation as each combination is categorised in the relationship types. In addition, other possibilities such as a tendency of rating increases in the dynamic course due to travel tendency are also categorised.

The parameter values are only categorised into high, low, positive, and negative values as well as values close to zero. The critical values for this classification are subsequently simulated in Section 8.3. Then exact values of the term "high" and "low" can be obtained.

The categorisation is done with the help of a flow chart, which can be seen in Fig. 7.10 and Fig. 7.11. The flow chart tests each parameter of the ASSG method in succession. Such a test of a parameter can then be satisfied or not satisfied. If it is satisfied, the flow chart can be followed by a green arrow with a "yes". If it is not satisfied, the flow chart must be continued by a red arrow with a "no". If the combination of the results of the parameters is unique enough, the flow chart will give a green box for a relationship type. The types for which no specific relationship can be found are categorised with a red "not related" box. This also marks the end of a the flow chart.

When the flow chart encounters a blue box, the parameter values are not unique enough to definitely lead to this type of relationship. However, it is possible that this type of relationship exists for the blue box. For further categorisation, the ASSG plot itself has to be examined. An example of this would be the possibility of a pseudo relation. Such a pseudo relation cannot be found unambiguously, but there may be signs in the parameters that a pseudo relation is possible. On the other hand, it is less likely that such a relationship type is present in the ASSG plot if no blue box has been gone through in the flow chart. In particular, if no blue box has been gone through, an examination of the ASSG plot itself is less important.

The interpretation through the travel tendency is indicated with grey boxes. There are three possibilities:

1. as time passes by the ratings increase ($tt > 0$),
2. as time passes by the ratings decrease ($tt < 0$),
3. as time passes the ratings do not tend in any direction ($tt \approx 0$).

Since there are many different parameters and relation types, the flow chart is divided into two pages. Therefore, the next page must be considered if the box "Go to next page" is fulfilled. Then one has to go from Fig. 7.10 to Fig. 7.11.

First, each categorisation starts at the top of the flow chart with the "Start" box. The first parameter that is tested is the travel distance. If the travel distance is very high, then the points in the ASSG jump around a lot, as in Fig. 7.7, which was a non-relational type. Therefore, if td is not low, the ASSG type is not relational and the flow chart is at the end.

If the td is low enough, the diagram continues with the next parameter. In this case, the inner point density is tested. Should the ipd be every high (e.g. 0.95), most of the points are within the deviation curve. Therefore, there could be a pseudo relation as in Fig. 7.4. This leads to the blue "may be pseudo related" box if the ipd is not low.

The χ^2 tests for the stationary relation follow. The categorisations for fixed x and y are the same. First, if the χ^2 value is close to zero, then the fit satisfies the χ^2 test and the indicators are stationary related. This leads to the green box "stationary related". Afterwards, the specific types for the three types of travel tendency are tested. This leads to the grey boxes for the χ_x^2 and χ_y^2 tests. If the χ^2 values are still low enough to have a stationary alignment, but high enough for some other type of relation, then the flow charts goes through a blue "may be stationary related" box. An example for such a case is Fig. 7.3, which also has a linear relation. Then the flow chart goes further down through the other parameters and possibilities. However, the flow chart also goes through the special types of driving tendency. This is one of the few cases of where the flow chart continues with two different arrows at the same time.

After the categorisation through the stationary types, the flow chart continues to the next page. Here the next tested parameter is the standard deviation difference. In the stationary type, the standard deviation difference is very high (see Section 6.6).

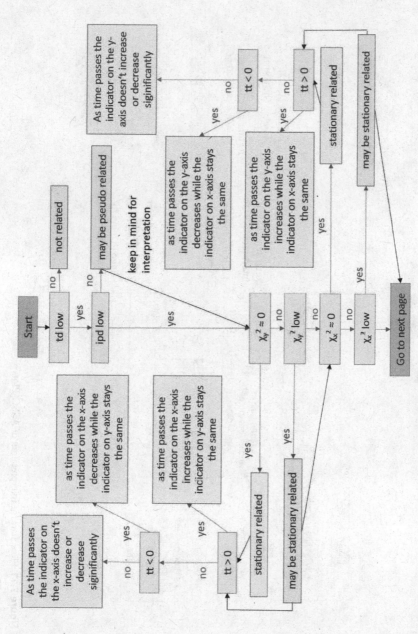

Figure 7.10 Flowchart for interpretation of ASSG-Plots

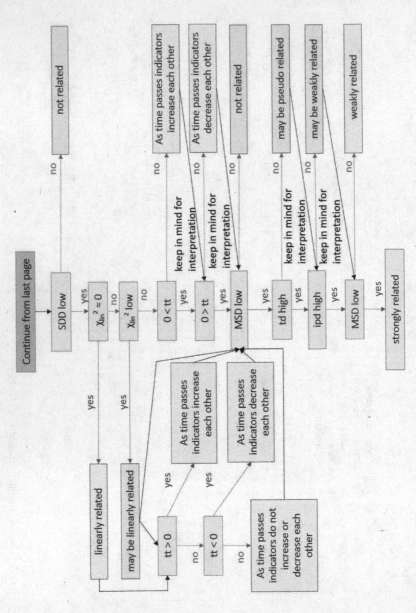

Figure 7.11 Flowchart for interpretation of ASSG-Plots

However, this type has already been covered before in this flow chart. Thus, if the standard deviation difference is still high at this point, then there is no other type of relation. Consequently, if the SDD is not low, the flowchart will result in a red "no relation" box. Otherwise, the standard deviation difference is low enough, the two indicators have a similar standard deviation and other types of relationship are possible.

The χ^2_{lin} test and the linear relation are then tested. This is analogous to the stationary type before. This again results in a test for the travel tendency and the distinction between the cases $\chi^2_{lin} \approx 0$ and χ^2_{lin} low.

Subsequently, the travel tendency is tested for the case that no χ^2 tests are low enough. At this point, both indicators must have a standard deviation that is already smaller than the critical value for the MSD of the field before. Therefore, a high absolute value of the tt leads to a dependence from one indicator to the other. This is different from the travel tendency test of the stationary types, where one indicator has to be at a fixed rating.

The last two boxes (tests of tt) are skipped if the flow diagram hits the green "linearly related" box or a blue "may be linear related" box. This is because the test of the travel tendency is already covered on the left-hand side. A second interpretation of the tt would then be unnecessary.

At this point it is not possible that only one indicator has a very high standard deviation and that the other indicator has a very low standard deviation. Since the SDD would be high in this case and the flow chart would have already resulted in "no relation". If the MSD is very high in this case, then the deviation curve is very large as in Fig. 7.8. In this example, the flow chart would have resulted in a blue "may be linear related" box and the red "no relation" box, because the MSD is too high.

The travel distance can be used to retest the ASSG in order to determine whether a pseudo relation is possible. If the travel distance is very low, then the points in the ASSG plot do not travel far. If the travel distance is low enough, then the points in the ASSG do not travel at all (see Fig. 7.4). Therefore, this part of the flow chart can retest whether a pseudo relation is possible. This means that it is now possible to hit a blue "may be pseudo related" box twice. If this is the case, it is more likely that a pseudo relation is present in the ASSG. However, if this box is only hit once, it is still valuable to examine the ASSG plot.

This leaves only two types of relationships: the weak relation and the strong relation. They differ in the size of the deviation curve and the number of points within the curve. Therefore, the two parameters corresponding to these two characteristics are tested. One of them is the inner point density, which is higher when the deviation curve is larger. Thus, if the ipd value is high enough, a strong correlation is more likely. However, it is also possible that the area of the deviation curve is so small that no point of the ASSG lies within the area. Such an example can be seen in Fig. 7.12.

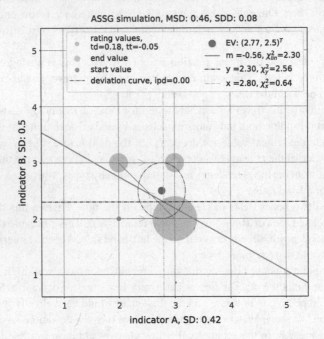

Figure 7.12 ASSG with $ipd = 0$

Nevertheless, in this case it is a strong relation type, even if the ipd is not high. Therefore, a lower ipd will only result in a blue "may be weakly related" box, as it could still be a strong relation. Additionally, the case of $ipd = 0$ will be included in the flow chart.

The other characteristic that remains to be examined is the size of the deviation curve. This size is estimated using the mean standard deviation. A low MSD means that the size of the deviation curve is small. This would result in a strong relation type in the green "strongly related" box. This means that it is also possible to pass through the blue "may be weakly related" box before finally ending up in the green "strongly related" box. For an interpretation, this suggests that the attractor attraction can be between a weakly related type and a strongly related type. This is due to the fact that the difference between these two types is defined by the MSD. Therefore, an ASSG with a strong relation and a blue "may be weakly related" box may not be as strong as an ASSG without such a blue box. However, since the difference between these two cases is very small, it is only relevant when trying to distinguish between different strong relations of indicators.

In contrast, a higher MSD signifies that the magnitude of the deviation curve is larger. However, at this point of the flow chart, the MSD is not too high to not have an attractor. This is due to the other test of the MSD before, which already ruled out this possibility. In contrast a higher MSD means that the size of the deviation curve is larger. Thus, the middle standard deviation is lower than for a no relation type, but too high for a strong relation. Thus, only the possibility of a weak relation remains, which leads to the green "weakly related" box.

Since there are only two green boxes left in the last test, the hover card is at the end. At this point, there must be at least one green or red box that has been hit. Therefore, this categorisation always leads to at least one category. Any further interpretation must then be done with the ASSG plot and the dynamic progression of its points.

7.3   Global Ratings with the ASSG Method

The last two chapters dealt with the interpretation of relation types between two indicators, which is one of the advantages of the ASSG method when compared to the SSG method. Another advantage over the SSG method are the global ratings of the ASSG with regard to the dynamic progression of the indicator ratings. This topic will be covered in this chapter.

For the interpretation of the course of an individual indicator, the parameters that estimate a characteristic of an indicator alone are relevant. This leaves only the standard deviation and the expected value of an indicator. In addition, the travel distance and the travel tendency of this single indicator can also be taken into account. These two parameters can be defined as follows:

Definition 7.3.1: Travel distance for a single indicator
Let $(x_i)_{i \in \{1,...,n\}}$ be the rating of an indicator x in the ASSG method. Then the travel distance of the indicator x in the ASSG method is:

$$td_x = \frac{1}{n-1} \sum_{i=1}^{n-1} |x_i - x_{i+1}|. \tag{7.1}$$

This definition is obtained by omitting the distance of a second indicator y from the definition of the travel distance of the ASSG plot in eq. 6.16. The same can be done with the travel tendency in eq. 7.6. In this case, the result of the sum has to be divided by $n - 1$ and not $2(n - 1)$ because the highest possible value of the sum in the travel tendency of one indicator is lower. The highest value is halved because half of the indicator functions are missing. Therefore, only $n - 1$ indicator functions can result in 1.

Definition 7.3.2: Travel tendency for a single indicator
Let $(x_i)_{i \in \mathcal{I}}$ be the rating of an indicator x in the ASSG method with $\mathcal{I} = \{1, ..., n\} \subset \mathbb{N}$. Then the travel tendency of the single indicator x in the ASSG method is

$$tt_x = \frac{1}{n-1} \sum_{i \in \mathcal{I}} \left(\mathbb{1}_{\{i \in \mathcal{I} | x_{i+1} - x_i > 0\}}(i) - \mathbb{1}_{\{i \in \mathcal{I} | x_{i+1} - x_i < 0\}}(i) \right). \tag{7.2}$$

The calculation of SD, tt, and td can be carried out with the rating of the indicators for each rater. In order to minimize the subjectivity of an individual rating, an average of all results of these parameters can be calculated for each rater. In addition, the standard deviation of the individual parameters provides information about the differences between the individual raters.

Corollary 7.3.1: Mean value and standard deviation of SD, tt, and td
Let $(a_i)_{i \in \{1, ..., m\}}$ be the result of the i-th rater of the parameter a, which is SD, tt, or td. Then the mean value of this parameter for all m raters is

$$\bar{a} = \frac{1}{m} \sum_{i=1}^{m} a_i \approx \mathbb{E}(a). \tag{7.3}$$

Additionally the standard deviation of this parameter a is

$$\sigma(a) = \sigma(a) \approx \sqrt{\frac{1}{m} \sum_{i=1}^{m} \left(\bar{a} - a_i\right)^2}. \tag{7.4}$$

Proof 7.3.1.: Mean value and standard deviation of SD, tt, and td
This corollary can be obtained by inserting the results of the ratings x_i and the total number of raters m into the corresponding equations of the standard deviations and expected value of a single indicator (see Section 6.2 and 6.3).   □

These single indicator parameters allow quantitative and thus also pictorial comparisons of the ratings of different individuals (see Figure 7.13a)–7.13d)). The mean of the expected value can also be calculated as above. The standard deviation of the mean of the expected value is by definition already the mean of the standard deviations. Therefore, the above definition only makes sense for the mean of the expected value.

Such mean values of the parameters can be considered as the final results of these parameters for the indicators assessed. How representative this result is can be estimated from their standard deviation. What this means can be explained with examples for different types of ratings in Fig. 7.13a)–7.13d).

In these plots, the rating of each rater is applied to the x-axis. Each rater has a fixed value on the y-axis. A blue dot on the graph represents a score at a fixed time interval for the rater on the y-axis and the score result on the x-axis. Similar to the ASSG plots, the size of each circle diameter at a point in the grid grows as it is hit more often. In addition, arrows between the points represent the dynamic progression of the points.

The red dot for each rater represents the expected value of the rater's ratings. The orange line is the size of the standard deviation of the ratings on either side of the expected value. This results in a symmetrical line at each expected value. The mean of the expected values is also visually plotted as a straight line with the fixed x-value at the result. The results of the means and standard deviations of the other three parameters can be found in the legend and on the x-axis.

The first Figure 7.13a) is an example of a group of scorers with very low standard deviations. In this case, most of the scores are located at one point. This location is also the result of the individual expected values. The expected values differ only slightly if there is a rating to the left or right of the concentration of most points. For example, the expected value of rater two is slightly further to the left than that of rater nine.

The standard deviations for raters with no rating other than 3 are zero. This can be seen because there is no orange line for the standard deviation around the expected value. For raters like rater two, where the ratings differ, a short line for the standard deviation can be found.

How large the standard deviation is on average can be read from the parameter $\overline{SD}$. In the case of Fig. 7.13a) this would be 0.15. How much this standard deviation deviates from the individual values can be estimated with SD_{SD}. For the first example, this deviation is almost as large as the mean value of the standard deviation itself. This means that this mean value does not represent the average standard deviations well.

In contrast, the mean of all expected values can be represented very well. This is due to the low mean value of the standard deviation. This can also be seen visually

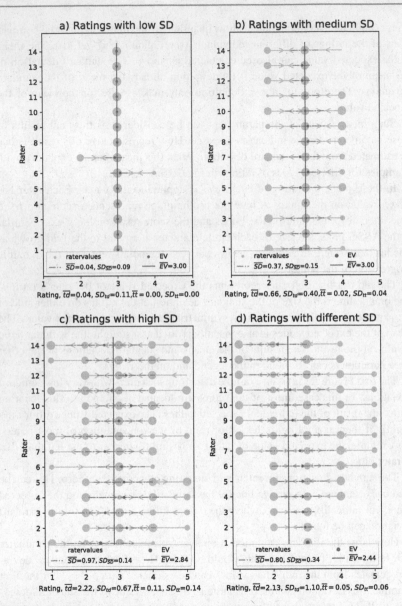

Figure 7.13 Ratings with different types of standard deviations

in the plot. Here the expectation values deviate only slightly from the red line of their mean.

How often the ratings jump can be estimated with the mean value of the travel distances. Since the ratings are mostly concentrated at the expected value, their travel distance is also very low. This also means that there is no dynamic time progression favouring one direction because they stay in the same place. The mean value of the travel tendency is thus zero.

The next example in Fig. 7.13b) shows a higher variability in the ratings. This is reflected in the higher mean value of the driving distances and the standard deviations. The mean of the expected values represents a lower overall score than in Fig. 7.13a) because the standard deviations are higher. This effect is even more pronounced in Fig. 7.13c), where the standard deviations are even higher. Here, the expected value does not represent an overall value at all. However, since each rating is gone through almost equally often, such a global value would not make sense anyway. This is also shown by the very high travel distance of 3.15, which can be represented for all ratings due to the low standard deviation $SD_{td} = 0.65$. The ratings therefore jump around too often for a single global value to represent a global tendency.

In contrast to the expected value, the mean of the standard deviations are easier to represent. This is due to the fact that the standard deviation of the mean of the standard deviations is comparatively smaller. This means that the standard deviations of the individual ratings are almost equal. A high and presentable standard deviation therefore means that an overall rating would not be presentable.

Different from this interpretation are the ratings in Fig. 7.13d). In this case, the mean of the standard deviation is high, which means that an overall value cannot be represented. However, the standard deviations also often differ because their standard deviation is high, which means that the standard deviation is also not representable for an interpretation. In such a case, the ratings of the raters differ so much that their results are hardly comparable. This could be caused by a very objective rating or insufficient training of the raters. Both cases would lead to each rater rating the indicator very differently. This means that the inter-coder-reliability is low (Kuckartz & Rädiker 2022). Thus, the data from these raters cannot be used for a global estimation of the indicator. Since the ASSG method can be used to obtain a measure of inter-coder reliability, it is not necessary to use other measurement tests such as the Cohen's kappa coefficient (eg. ebd.).

In sum, $\overline{EV}$ a representable global value if the standard deviations are low. If $\overline{SD}$ is high and SD_{SD} is comparatively low, then $\overline{SD}$ is a representable global value. However, $\overline{EV}$ can represent a global outcome of the indicator, while $\overline{SD}$ only represents how much the outcomes of the ratings differ through the time intervals.

7.4   Multistabilities and Disruption Intervention

As explained in Section 7.1.6, the ASSG method struggles with more than one stability, so called multistabilties (see Section 4.2.1). This chapter presents a possible solution and the implications for the interpretations.

Imagine two indicators that rate different aspects of the basic dimension of student support. In the first few minutes, a disturbance occurs that changes the scores to results close to $(1, 1)^T$. Such a disturbance could for example be when a student starts an argument with another student. Other, less extreme examples would be a verbal argument between a student and a teacher or bullying situations among students. Then the teacher needs some time to resolve the situation, which leads to a stability in the situation of $(1, 1)^T$. After the situation is resolved, the student-teacher system reaches a better state, leading to a stability in $(5, 5)^T$. In the ASSG method, a plot for such an example would look like Fig. 7.14.

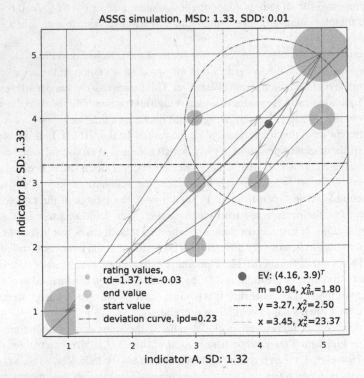

Figure 7.14 ASSG for two stabilities

The described stabilities at $(1, 1)^T$ and $(5, 5)^T$ can be seen in the larger points at their locations. This is the same ASSG that was already discussed in Section 7.1.6. In this kind of ASSG plot, an estimate of global variability and a single attractor cannot describe an ASSG with two stabilities.

However, these two stabilities are only after each other. Therefore, one can describe the first stability at $(1, 1)^T$ with the parameters of the ASSG method using only the first half of the rankings. The remaining half for the second stability can then be estimated with a new set of parameters. This would result in an ASSG plot with two expected values and two deviation curves. For the example in Fig. 7.14, this idea would result in an ASSG plot as in Fig. 7.15.

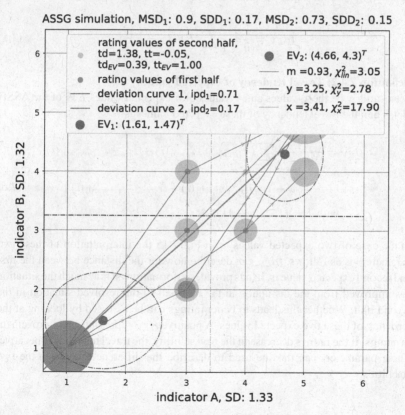

Figure 7.15 ASSG with two deviation curves and expected values

In this case, the expected values can estimate the position of the two attractors. The deviation curves and the mean standard deviations describe the variability of the individual attractors. With these two sets of parameters, each stability can be described separately. These two stabilities can be classified into the different types from Section 7.1.

In addition, the travel distance (td_{EV}) and travel tendency (tt_{EV}) between the expected values can provide further insight. These parameters are defined analogue to their counterparts of the indicator ratings (see Section 6.8 and Section 6.10).

Definition 7.4.1: Travel distance of expected values

Let $(EV_i)_{i \in \mathbb{N}}$ be the expected values of n attractors. Then td_{EV} is defined as

$$td_{EV} = \frac{1}{n-1} \sum_{i=1}^{n} |EV_{i+1} - EV_i|. \tag{7.5}$$

Definition 7.4.2: Travel tendency of expected values

Let x_i and y_i be the x-values and y-values of n expected values EV_i of the ASSG plot. Then the travel tendency of these expected values is

$$tt_{EV} = \frac{1}{2(n-1)} \sum_{i \in \mathcal{I}} \Big(\mathbb{1}_{\{i \in \mathcal{I} | x_{i+1} - x_i > 0\}}(i) + \mathbb{1}_{\{i \in \mathcal{I} | y_{i+1} - y_i > 0\}}(i)$$
$$- \mathbb{1}_{\{i \in \mathcal{I} | x_{i+1} - x_i < 0\}}(i) - \mathbb{1}_{\{i \in \mathcal{I} | y_{i+1} - y_i < 0\}}(i) \Big) \tag{7.6}$$

with $\mathcal{I} = \{1, ..., n-1\} \subset \mathbb{N}$.

In the case of two expected values as in Fig. 7.15 the interpretation of these two parameters is as follows. td_{EV} can describe how far the distance between the first and second expected value is. It thus provides information on how much the situation has improved from the disruption in the first half to the resolved situation in the second half. Whether this leads to better ratings can be assessed by looking at the tendency of these two expected values. A positive tt_{EV} represents an improvement in ratings. If the ratings decrease in the new stability, the travel tendency is negative. These parameters can thus be used to describe the differences between the two stabilities.

With this idea, it is now possible to describe multistabilities if the time intervals for each stability are known. It is necessary to look at the assessments of the indicators in detail to distinguish between different stabilities. The ASSG method as it stands cannot distinguish between one or more stabilities without individual input from the researcher.

The last example of an implication is very specific and can only be applied in a situation that starts immediately with a long disturbance. A less rare example would be when the student-teacher situation is initially stable in an attractor. Then an rating occurs outside the stability. After this is resolved, the student-teacher system is in stability in another attractor. Such an ASSG can be seen in Fig. 7.16.

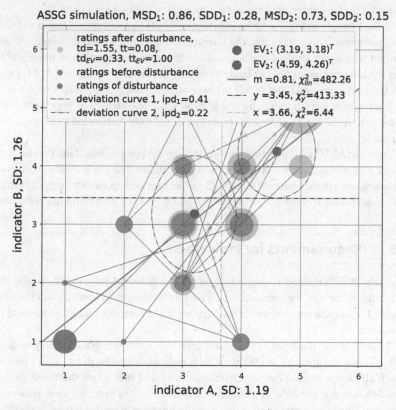

Figure 7.16 ASSG for disruption intervention

Here the points before the interruption are in a darker blue than the points after the interruption. The evaluations of the interruption itself are marked with red coloured points. The two stabilities around $(3, 3)^T$ and $(4.5, 4)^T$ are estimated with the two sets of parameters as in the last example. In this case, the points of the interruption are not described by their own parameter set because the duration of the interruption is too short.

With the difference of the parameters, further insight into the intervention of the disturbance is given. Since the student-teacher system is only in the state of disturbance for a short time, something must have happened that led to the stability in $(4.5, 4)^T$. This can for example be when the teacher resolves the situation, as in the last example, or the situation resolves itself. In the case of an intervention, the after-effects of the intervention can be estimated with the ASSG method.

In the example in Fig. 7.16, the intervention leads to a better ranking in the new expected value. This means that the intervention not only solved the situation, but also led to a better student-teacher system. Whether the situation is more stable after the intervention or not can be described by the difference between MSD_1 and MSD_2. In this example, the MSD is lower after the intervention. This means that the intervention has led to a more stable pupil-teacher system. The opposite would be the case if the MSD decreases. In this case, the intervention led to a less stable situation.

Thus, the ASSG method can therefore not only describe different interventions of disorders, but also the after-effects of the interventions. These after-effects can also be ranked, as each parameter of the ASSG method has a comparable numerical value. This would make it possible to compare different types of disruption interventions.

7.5   Requirements for Indicators

Since the ASSG method is dependent on the weak law of large numbers and the variability of ratings, the indicators used for this method must meet certain requirements. These requirements and what happens if they are not met are discussed in this chapter.

An indicator in the ASSG method has to be able to show a dynamic progression. If this is not possible, an ASSG with such indicators results in an ASSG plot in which all ratings remain the same. With such a set of indicators, therefore, only a pseudo-relation is possible. A global rating is still possible in such a case. However, there can hardly be a meaningful result of the interaction between these indicators (see Section 7.1.3).

Such dynamic progression is not possible if, for example, the occurrence of higher and lower rankings is very unlikely. Therefore, the difference in ranking must be achievable for a typical student-teacher system.

A second requirement is that the indicator has more than three possibilities in the evaluation system. An indicator that differs only between two states cannot have a high variability in the dynamic progression. An ASSG with two indicators with only two possible outcomes can only lead to four possible locations in the ASSG plot (see Fig. 7.17). In contrast, an ASSG with indicators with five possible outputs can lead to 25 different locations. Thus, an indicator with more possible ratings can have much larger differences in variabilities. A too small number of possible locations may result in no distinction between weak and low variability. This is because the ASSG plot can only look like a pseudo relation (see Fig. 7.18), a strong relation (Fig. 7.19), or an unrelated case (Fig. 7.20). Any other type of variability between these types cannot be represented due to the lack of possible locations.

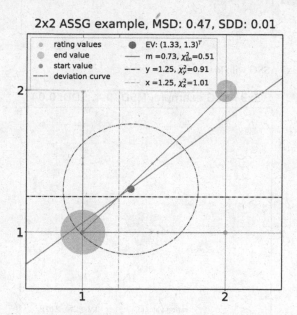

Figure 7.17 Example for a 2x2 ASSG

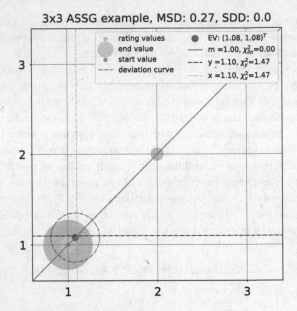

Figure 7.18 3x3 ASSG with low variability

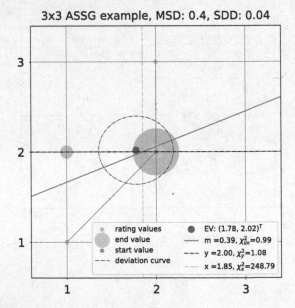

Figure 7.19 3x3 ASSG with medium variability

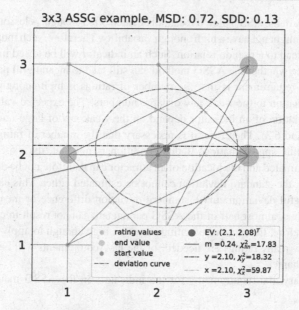

Figure 7.20 3x3 ASSG with high variability

Therefore, at least four different rankings are needed to distinguish between weak and strong relations. A higher number of locations increases the possible variabilities and reduces the probability of a pseudo relation. However, the categorisation of relation types in Section 7.1 is optimised for indicators with five possible rankings. The simulations in the following chapter are also only carried out for indicators with five rankings. Indicators with five rankings are therefore recommended for the ASSG method.

Another requirement for the indicators is that they have to be repeatable and rateable over any given time interval. An indicator that only describes a global aspect of the teacher-student system cannot have a dynamic progression. For example, an indicator that describes how the teacher starts the lesson can only have one rating over the duration of the lesson. Such an indicator can have five different scores, but dynamic progression is not possible. An ASSG with this indicator can not be meaningful because there is only one point in the ASSG plot.

Nevertheless, an indicator describing how a teacher starts a lesson section could be used for an ASSG plot. For ratings that are done every two minutes, this indicator cannot be rated every single time. To solve this problem, the rating of the previous assessment can be used. In this way, such an indicator could be paired with other indicators that can be scored more frequently. However, such an indicator decreases the variability of the ASSG plot, if it cannot be rated often enough. A lesson with only two sections might not have a high enough variability. Therefore, such indicators can more easily lead to a pseudo relation. Such an indicator will be tested in Chapter 9 in order to see whether the ASSG method can still lead to meaningful parameters.

The final requirement is that the number of ratings is high enough to satisfy the approximation to the weak law of large numbers. The expected value and the standard deviation of an indicator depend on the weak law of large numbers (see Section 6.2 and 6.3). Therefore, it is necessary that the number of ratings is large enough for this approximation. If this condition is not met, the expected value cannot be estimated and the location of an attractor can therefore not be determined. Furthermore, the standard deviation cannot be estimated either. This can not lead to a meaningful deviation curve, standard deviation difference or mean standard deviation. Thus, almost half of the ASSG parameters can not result in comparable values. Therefore, the number of ratings must be high enough to apply the ASSG method. How many ratings are needed for the last requirements will be simulated in the next chapter.

In summary, the requirements for the indicators in the ASSG method are as follows:

1. The indicator ratings are achievable for a typical teacher-class system.
2. The indicator has at least four possible scores, however, five are recommended.
3. The indicator is repeatedly rateable.
4. The indicator has sufficient ratings to approximate the weak law of large numbers.

These requirements severely limit the number of indicators that can be used. For example, indicators that have the outcomes "yes" or "no" can not be used. Indicators that rate aspects of individual time events, such as the beginning or end of the lesson, are also not usable. This leaves the researcher with only certain types of indicators and makes the search for a faulty indicator even more time-consuming than before. However, the advantages of the ASSG method can outweigh such a disadvantage, as a dynamic progression cannot be analysed with the other research methods from 3.

In addition, more exotic indicators such as the eye tracking indicator (Zhao et al. 2014) can also be used in the ASSG method. Such indicators can be measured frequently enough and can have a high variability. Another type of usable indicators are physically measurable aspects, such as the loudness of background noise (Peng et al. 2018). These indicators can be measured repeatably with separate devices and do not require a researcher to rate these indicators.

Simulation of global ratings and relation types

8

In this chapter, the two outstanding problems are to be solved. The first problem is how many ratings are needed for the approximation with the weak law of large numbers. The second problem is to find the critical values of the flow chart for categorising the relation types. When this problem is solved, the research question "How can different types of SSG plots be categorized?" can be answered.

Both problems are solved using simulations of indicator ratings. The ratings of indicators in an ASSG plot are simulated by using random variables of distributions or stochastic processes as ratings on the axes of the ASSG plot. With this concept, different combinations of distributions and stochastic processes can be used for different cases in an ASSG plot. These simulations will be carried out with random variables from Section 5.3. How the relation types can be simulated with these random variables is discussed in Section 8.1. Subsequently, the applicability of the approximation with the weak law of large numbers is simulated in Section 8.2. Subsequently, the critical values for the flow chart from the previous chapter (see Fig. 7.10 and Fig. 7.11) will be simulated in Section 8.3.

8.1 Simulating Ratings with Random Variables

In this chapter, the combinations of random variables that still count as a relation type are determined. There is no empirical possibility of comparison, as the relation types were determined by the alignment of the points in the ASSG plot. Therefore, in this chapter the combinations for each relation type were determined by looking at many different relations of ASSG plots for these combinations.

The first type of relationship that can be simulated is the pseudo relation. This type of relation requires two indicator ratings with low variability. Such ratings can be simulated with a Gaussian distribution with low standard deviation σ. The parameter μ of the Gaussian distribution is also the expected value of the ratings on one axis (see proof 5.4.3).

An example of a simulated ASSG plot of Gaussian distributed indicator ratings with low variance can be seen in Fig. 8.1. In this case, each result of the random variable is rounded to assign it to the grid. This results in almost all ratings being 3. Only a small proportion of the ratings are not rounded to 3. However, the probability of such results is very low because the variance of the ratings is small. Therefore, a larger deviation from the expected value μ is very unlikely. This leads to the many points at $(4, 3)^T$ or $(3, 4)^T$ and a small circle diameter at these locations. Thus, most of the ratings are located at the attractor and only a few are not. This leads to a simulation of a pseudo relation or a strong relation with very little variability.

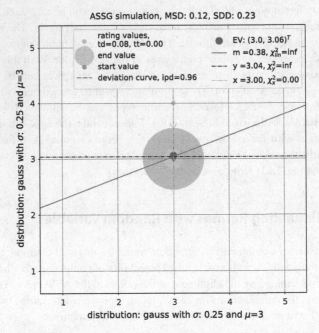

Figure 8.1 ASSG example of two Gaussian distributed indicator ratings with low variance

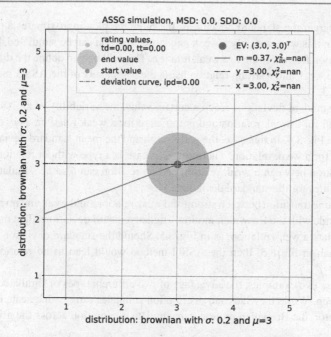

Figure 8.2 ASSG example of two Brownian motions with low variance

An alternative to the Gaussian distributions are rounded Brownian motions with small standard deviations in each Gaussian decrement. Such an ASSG plot with indicator ratings following a Brownian motion can be seen in Fig. 8.2. The starting value of this stochastic process is also called μ. This is because the next chapter will show that the starting value is also the expected value of the ASSG method if the standard deviation of the Gaussian distributed decrements is small enough. In the case of Fig. 8.2, therefore, all the ratings are located at $(3, 3)^T$. In this case, the variance of the decrements is so small that all ratings are rounded to 3 and no other rating occurs. Thus, a pseudo relation or a strong relation without variability is simulated here as well.

A weak relation can also be simulated with Gaussian distributions or the use of Brownian motions. For this, the standard deviations of the Gaussian distributions must be higher, as in Fig. 8.3. With higher variability, values other than μ can occur with higher probability. This still leads to a strong attractor at the location of the expected value of the ASSG, but also to some other locations with a larger circle diameter in the vicinity.

The standard deviation can be used to simulate how far the ratings deviate around an attractor, as smaller standard deviations are more likely to result in a plot like

8.1. The larger the standard deviations, the more the plot approximates a plot like Fig. 8.3. In this way, not only a weak and strong relation can be simulated, but also the transition from a strong to a weak relation. This will help to define the difference between a weak and a strong relation using the parameters of the ASSG method in Section 8.3.

If the standard deviations reach a certain value, the variability of an ASSG plot is too high for a weak relation and is too large for a weak relation. This case can be seen in Fig. 8.4. In this case, the path length and the mean standard deviation are too large for a weak relation. Therefore, this case is a type without relation. Thus, the transition between a weak relation and no relation can also be simulated with higher values for the standard deviations.

The same transition between strong and weak relations and weak and no relations can be made with the Brownian motion. Higher standard deviations than in Fig. 8.2 can simulate a weak relation, as in Fig. 8.5. Should the standard deviation be even higher than in Fig. 8.5, then the ASSG method would lead to no relation, as in Fig. 8.6.

In these two examples, the advantage of two different types of simulated ratings can be seen. While the Gaussian distribution simulates ratings that scatter around the attractor, the Brownian motion can simulate a migration across the grid in the

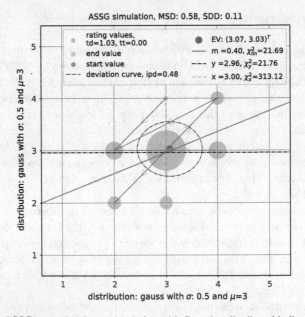

Figure 8.3 ASSG example of a weak relation with Gaussian distributed indicator ratings

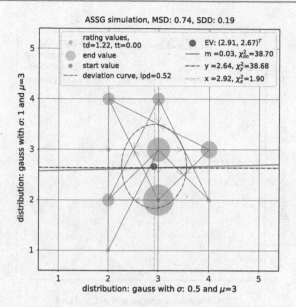

Figure 8.4 ASSG example with too high standard deviations for a weak relation

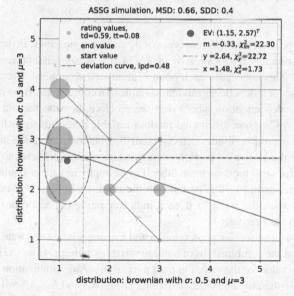

Figure 8.5 ASSG example of a weak relation with Brownian motions

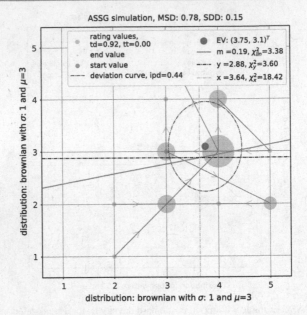

Figure 8.6 ASSG example too high variability for a weak relation with Brownian motions

ASSG plot. These are two different types of dynamic progressions. In the case of the Gaussian distribution, the location of a point in time is independent from any other point in time. This is because each point is simulated separately. Therefore, these ratings are a set of independent random variables. In contrast, the Brownian motion valuations are dependent on where they were before, because the old location is added to a new Gaussian distributed random variable that leads to each decrement of the Brownian motion. While each Gaussian distributed decrement is independent, the ratings of the Brownian motion are not independent. This leads to a dynamic progression from one location to an other. For example the ratings could tend more to the left side, as in Fig. 8.5, or more to the right, as in Fig. 8.6. Such dynamic progressions would be possible if each indicator rating depended heavily on the situation of the last rating.

To distinguish between a weak relation and a strong relation and a strong relation and no relation, the combinations of random variables in the previous examples serve as a boundary value for the values of their parameters. Any combination of standard deviations less than the standard deviations in Fig. 8.3 and Fig. 8.5 will be treated as a strong relation. Combinations of standard deviations less than or equal to the standard deviations in Fig. 8.1 or Fig. 8.2 will be treated as strong relations with a high possibility of a pseudo relation. Should a combination of standard deviations be between Fig. 8.3 and Fig. 8.4 or between Fig. 8.5 and Fig. 8.6 they will be treated as weak relations. Any other combination with higher standard deviations will be treated as an ASSG without a relation.

This differentiation was done by examining the alignment of the points in the ASSG plots for many realisations of these combinations of random variables. The above figures only serve as examples for many other ASSG plots with the same combination of random variables. Since more than 3000 ASSG plots were used to determine the boundaries for the relation types, these ASSG plots are not presented in this thesis. A presentation of all these realisations would consume too much space and would not contribute to a better understanding. Therefore, only the examples that represent the typical arrangement of these combinations for these random variables are presented in this chapter.

This leaves the stationary relation types and the linear relation. Since the two stationary relations only differ in the roles of the two indicators, the simulation of the parameters can only be carried out for one of these types. For the flow chart in Section 7.2, only the critical values of the χ^2 test have to be simulated. A simulation of the linear relation type is also not necessary. This is because the alignment of the points differs only by the angle of the constant to which the points are aligned in the ASSG. For example, a linear relation with $m = 1$, differs by an angle of 45 degrees from a stationary relation with a fixed y-value. Thus, only one of these relation types needs to be simulated.

One way to simulate a stationary relation is through the use of the discrete uniform distribution. A combination of two ratings with the discrete uniform distribution would simulate an ASSG plot where no relation should be possible in the flow chart. This combination is also the case with the greatest possible variability in the ASSG method. Therefore, this combination can serve as an upper boundary for parameters that depend only on the variability in the ASSG plot. Such an ASSG plot can be seen in Fig. 8.7.

By using a combination of a uniform distributed indicator rating and a Gaussian distributed indicator rating with a very low standard deviation, a stationary related ASSG plot can be simulated. An example of this combination can be seen in Fig. 8.8. This combination simulates a stationary relation with a fixed x-axis serves as a test for the critical values of the χ^2 values. With these three types of random variables, the parameters of all relation types can thus be simulated, with the exception of the linear relation, which is not needed.

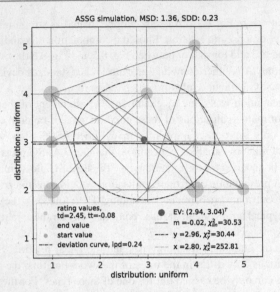

Figure 8.7 ASSG example of two uniform distributed indicator ratings

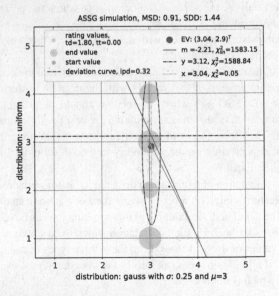

Figure 8.8 ASSG example for a stationary relation

In summary the critical values of every relation type can be simulated with the following combinations:

Table 8.1 Combinations of distributions for boundaries in the relation types

random variable on x-axis	random variable on y-axis	boundary for relation type
Gaussian, $\sigma = 0.25$	Gaussian, $\sigma = 0.25$	lower are pseudo related
Brownian, $\sigma = 0.2$	Brownian, $\sigma = 0.2$	lower are pseudo related
Gaussian, $\sigma = 0.5$	Gaussian, $\sigma = 0.5$	lower are strong related
Brownian, $\sigma = 0.5$	Brownian, $\sigma = 0.5$	lower are strong related
Gaussian, $\sigma = 0.5$	Gaussian, $\sigma = 1$	lower are weak related
Brownian, $\sigma = 1$	Brownian, $\sigma = 1$	lower are weak related
uniform	uniform	boundary for no relation
Gaussian, $\sigma = 0.25$	uniform	stationary related

However, in order to calculate the expected value of the ASSG, it is necessary to apply the weak law of large numbers to approximate the expected value. This means that the set of random variables used for the indicator ratings must satisfy this approximation when required. This is the case for strong and weak relations where a single indicator exists. Whether this approximation is possible is simulated in the next chapter.

8.2   Expected Value – Weak Law of Large Numbers

The aim of this chapter is to answer the still open question of how many ratings the ASSG needs in order to be able to use the weak law of large numbers for the approximation of the expected values. In order to assess the applicability of this theorem, the effect of the weak law of large numbers for the different random variables from Section 8.1 is examined. The effect to be investigated is that histograms of mean values have to be concentrated at the expected value (see Section 5.6). If the histogram is slim enough, then the approximation can be used. This is because results outside the expected value are unlikely if most of the results are at the expected value. Therefore, it is very likely that the expected value can be approximated by the mean for an ASSG plot.

First, this effect is examined with Gaussian distributed indicator ratings. For the histogram, 100 simulations of the mean value are calculated. Each mean value is calculated with n Gaussian distributed indicator ratings. In the case of Fig. 8.9, each mean value is calculated with $n = 20$ Gaussian distributed indicator ratings with $\mu = 3$ and $\sigma = 0.25$. The expected value where most of the results of the histogram should be would be $\mu = 3$. This is because μ is the expected value of the Gaussian distribution (see Section 5.4).

In the case of Fig. 8.9, 70% of the calculated mean values are between 2.95 and 3.05. Whereas the remaining 30% of the results are between 2.85 and 2.95 or 3.05 and 3.15. This means that about 70% of the mean value results approximate the expected value with an error of ±0.05. In addition, all mean values approximate the expected value with an error of ±0.15.

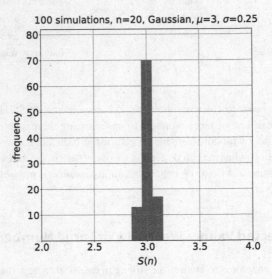

Figure 8.9 Histogram: 100 simulations of mean values of 20 Gaussian distributed random variables with $\sigma = 0.25$ and $\mu = 3$

A similar histogram can be produced with $n = 10$ ratings for each mean value. This case is shown in Fig. 8.10. Here the probability of staying between 2.95 and 3.05 decreases by 15%. Additionally there are a few results that fall outside the error of ±0.15. The probability of obtaining a result with an error of ±0.15 also increases significantly. This means that the estimate of the expected value and thus the location of an attractor is better with twenty indicator ratings than with ten indicator ratings for the case of a pseudo relation. However, it is very likely in both cases that the

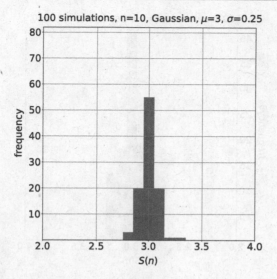

Figure 8.10 Histogram: 100 simulations of mean values of 10 Gaussian distributed random variables with $\sigma = 0.25$ and $\mu = 3$

mean value is rounded to the correct result of a global indicator rating. Only the error of the first decimal place increases with lower indicator ratings.

The histogram becomes less narrow with a standard deviation of $\sigma = 0.5$. Here, even with $n = 20$ number of indicator ratings, 18% of the results are outside of an error of ±0.15 (see Fig. 8.11). The approximation gets worse with $n = 10$. In Fig. 8.12 even fewer results are within a corridor with an error of ±0.15. Nevertheless, it is still possible to round the result of the mean value to a correct expected value, as all results still have an error less than ±0.5.

However, with even higher standard deviations such as $\sigma = 1$, 3% of the outcomes are outside an error of ±0.5 (see Fig. 8.13). These results occur even more frequently with a lower number of indicator ratings (see Fig. 8.14), where only 89% of the results have an error less than an ±0.5. This means that with a standard deviation of $\sigma = 1$ there is a small probability that the means of $n = 20$ indicator ratings do not correctly estimate a global expected value when rounded. This probability increases significantly (from 3% to 8%) when there are only $n = 10$ ratings. Thus, only ten indicator ratings lead to a mean value that is incorrect after rounding with a probability of roughly 10%, for the case of Gaussian distributed indicator ratings with $\sigma = 1$. With twenty indicator ratings, the probability of this error is more than halved.

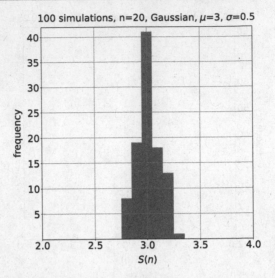

Figure 8.11 Histogram: 100 simulations of mean values of 20 Gaussian distributed random variables with $\sigma = 0.5$ and $\mu = 3$

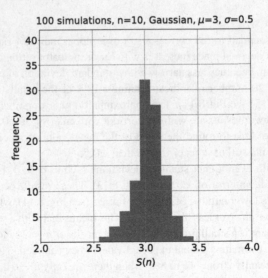

Figure 8.12 Histogram: 100 simulations of mean values of 10 Gaussian distributed random variables with $\sigma = 0.5$ and $\mu = 3$

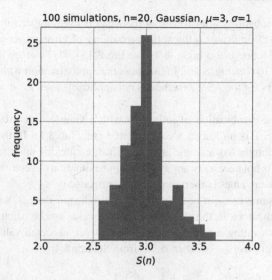

Figure 8.13 Histogram: 100 simulations of mean values of 20 Gaussian distributed random variables with $\sigma = 1$ and $\mu = 3$

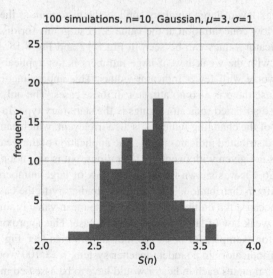

Figure 8.14 Histogram: 100 simulations of mean values of 10 Gaussian distributed random variables with $\sigma = 1$ and $\mu = 3$

This error probability after rounding the mean is even higher for $\sigma = 2$. Here, even with $n = 20$ indicator ratings, the probability of error for an approximation to the expected value is too high 18% (see Fig. 8.15). This chance increases even more with $n = 10$ (see Fig. 8.16). In this case, the approximation with the weak law of large numbers for the ASSG method is not applicable even for $n = 20$ indicator ratings.

However, the combination of indicators with a Gaussian-distributed indicator rating and $\sigma \geq 1$ is no longer weakly related (see Tab. 8.1). In these cases, the ASSG plot no longer has a single attractor. Thus, estimating an attractor with the expected value is not necessary anyway. The approximation of such expected values by means of mean values is therefore no longer necessary.

This is the same case for equally distributed indicator ratings, which also lead to unrelated indicators. In this case, the expected value can be calculated with the highest indicator rating result $m = 5$ and the proven expected value for discrete uniform distributions (see equation 5.18) from Section 5.4:

$$\frac{m+1}{2} = \frac{5+1}{2} = 3.$$

Thus, if the weak law of large numbers applies, the histogram of the mean values should show a low concentration at the value 3. Histograms for discrete, equally distributed indicator values can be seen in Fig. 8.17 and Fig. 8.18. However, the approximation with the weak law of large numbers is not applicable here either and becomes worse with fewer indicator values. This approximation is also not necessary because there is also no attractor in these cases. The only relation type with uniformly distributed indicator ratings is the stationary type. In this case, the expected value of the changing indicator is also irrelevant, which can be simulated by a uniformly distributed indicator rating. The application of the weak law of large numbers is thus not given with twenty or ten ratings, but is also not needed.

In Section 5.6.1 it was shown that the weak law of large numbers can be used for discrete uniform distribution. This is not contradictory to the case in Fig. 8.17 and Fig. 8.18, because the number of values for the mean value is simply not large enough for the weak law of large numbers in this case. This approximation would be applicable, for example, for $n = 100$ indicator ratings (see Fig. 5.15). For an actual rating of indicators in a student-teacher system, $n = 100$ would mean that in a lesson of 45 minutes each indicator would have to be assessed more than once every 30 seconds. Such a high frequency of assessment is not easy to achieve. So far, it is still possible to approximate the expected value with an error of ± 0.5 using only $n = 20$ indicator ratings, for those cases where the approximation is needed.

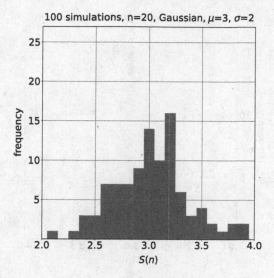

Figure 8.15 Histogram: 100 simulations of mean values of 20 Gaussian distributed random variables with $\sigma = 2$ and $\mu = 3$

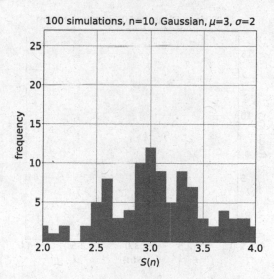

Figure 8.16 Histogram: 100 simulations of mean values of 10 Gaussian distributed random variables with $\sigma = 2$ and $\mu = 3$

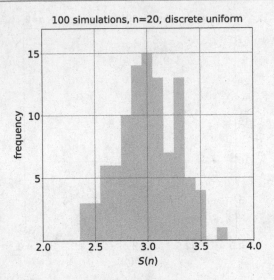

Figure 8.17 Histogram: 100 simulations of mean values of 20 uniform distributed random variables

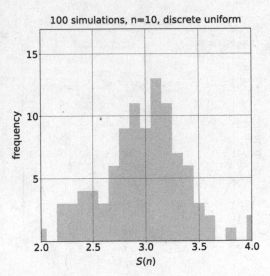

Figure 8.18 Histogram: 100 simulations of mean values of 10 uniform distributed random variables

This means that one rating every 2 minutes in an 45 minute long lesson is sufficient for the ASSG method.

The situation is different for indicator ratings that follow a Brownian motion with small standard deviations. With Brownian motions, the indicator ratings are no longer independent, as each result of each step depends on the last result. Therefore, the weak law of large numbers is not applicable as the premise is not fulfilled. This can also be seen in the histograms of Fig. 8.19. and Fig. 8.20. In these cases the histograms are scattered, which means that the mean value does not converge. The approximation to the mean is therefore not sufficient for the ASSG method.

However, the Brownian motion has results that are not based on natural numbers. Thus, each result of the Brownian motion has to be rounded. The use of such rounded indicator values leads to histograms as in Fig. 8.21 and Fig. 8.22. In these cases, almost all results of the mean lie within an error of ± 0.05 even with only $n = 10$ ratings. As mentioned in the previous Section 8.1, the attractor is located at the starting value of the Brownian motion, if the standard deviation is low enough. Since both histograms are located at this starting value of $x = 2$, the approximation of the mean for the strong relation type is applicable in an ASSG plot.

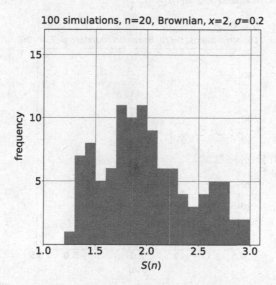

Figure 8.19 Histogram: 100 simulations of mean values of 20 outcomes of Brownian motions with the start value $x = 2$ and $\sigma = 0.2$

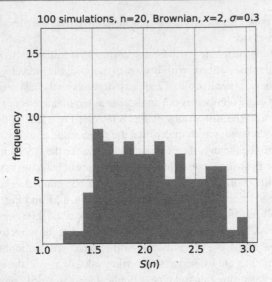

Figure 8.20 Histogram: 100 simulations of mean values of 20 outcomes of Brownian motions with the start value $x = 2$ and $\sigma = 0.3$

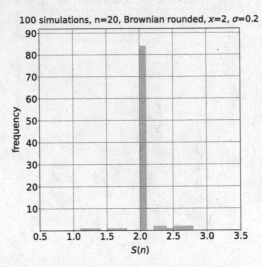

Figure 8.21 Histogram: 100 simulations of mean values of 20 rounded outcomes of Brownian motions with start value $x = 2$ and $\sigma = 0.2$

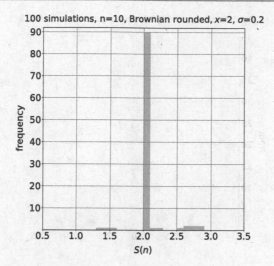

Figure 8.22 Histogram: 100 simulations of mean values of 10 rounded outcomes of Brownian motions with start value $x = 2$ and $\sigma = 0.2$

This high peak in the histogram remains similar even with only $n = 10$ ratings and a higher standard deviation of $\sigma = 0.3$ (see Fig. 8.23). Nevertheless, there are some outcomes with a higher error than ± 0.5. With an even higher standard deviation of $\sigma = 0.5$, the peak at $x = 2$ disappears, as in Fig. 8.24. Here the histogram is more scattered than with the evenly distributed indicator ratings in Fig. 8.18. This means that even the mean values of the rounded Brownian motion results do not converge to the same result in the case of weak relations.

To further clarify why the means do not converge, several results from the simulated ASSG plot are helpful. Fig. 8.25 and Fig. 8.26 both show the same combination of 25 indicator ratings following a Brownian motion with a starting value of $\mu = 3$ and standard deviation of $\sigma = 0.5$.

Although the indicator ratings all follow the same Brownian motion, the results are very different. The attractor of the left ASSG plot (Fig. 8.25) is in the lower left corner of the plot, while the attractor in the ASSG plot (Fig. 8.26) on the right side is in the upper right corner. Nevertheless, the expected values approximated by the mean values still show the correct estimation of the location of the attractors. Their results are just very different. Thus, a histogram of different indicator assessments of the same Brownian motion leads to many different mean values. This leads to a more spread out histogram as in 8.24, because the mean values are not equal.

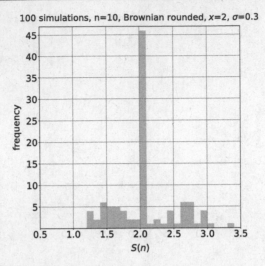

Figure 8.23 Histogram: 100 simulations of mean values of 10 rounded outcomes of Brownian motions with start value $x = 2$ and $\sigma = 0.3$

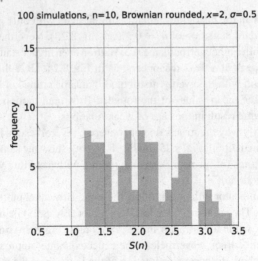

Figure 8.24 Histogram: 100 simulations of mean values of 10 rounded outcomes of Brownian motions with start value $x = 2$ and $\sigma = 0.5$

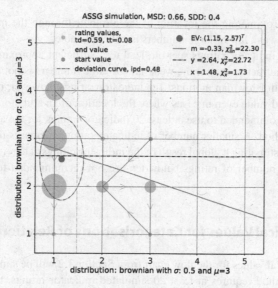

Figure 8.25 First ASSG example with two indicators ratings following a Brownian motion with start value $\mu = 3$ and $\sigma = 0.5$

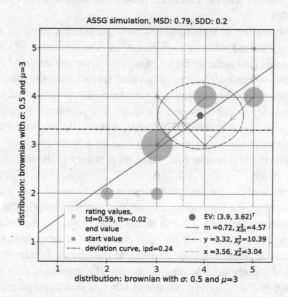

Figure 8.26 Second ASSG example with two indicators ratings following a Brownian motion with start value $\mu = 3$ and $\sigma = 0.5$

Therefore, approximating the location of the attractor through the mean value is useful, but the weak law of large numbers is not applicable.

In total, 20 indicator ratings seem to lead to a sufficient approximation of the location of an attractor, even if the weak law of large numbers is not applicable to simulations with Brownian motions. Ten indicator ratings, however, may lead to a wrong expected value even in cases where the location of an attractor is needed. It is therefore recommended to use at least 20 indicator ratings for the application of the ASSG method. A smaller number of indicator ratings increases the likelihood of an error in estimating a global result of the indicator ratings. Therefore, indicator ratings with a number of ratings below ten are not recommended for the ASSG method.

8.3   Critical Values for Categorisations of Relation Types

Next, the critical values for the flow chart from Section 7.2 will be simulated. Since the ASSG method requires at least 20 simulated indicator ratings, the following simulations will use 25 indicator ratings. With 25 indicator ratings, each parameter of the ASSG method will be simulated 10000 times for a fixed combination of random variables.

Due to the strong law of large numbers, these parameters converge and result in a representative expected value for each parameter. The strong law of large numbers is applicable in this case because the set of simulated parameters can be interpreted as a set of random variables. These random variables then have the same distribution because the indicator ratings leading to these parameters are all equally distributed. The results are independent because each result of a parameter is calculated from different independently simulated ASSG plots. Finally, the results for each parameter are finite, except the χ^2 values, which lead to a finite expected value of the parameter results to the power of four. Thus, all premises of the strong law of large numbers are fulfilled.

The application of the strong law of large numbers means that in this case, as long as the results of the parameters are simulated often enough, the mean value of these parameters converges towards the expected value of the parameter. This means that a representative result for each parameter of any combination of the distributions can be simulated with a sufficient number of simulated parameters. To achieve a sufficiently high number of simulated parameters, each parameter is simulated 10000 times. Verifying that this number of simulated parameters is high enough will be investigated later on in this chapter.

Table 8.2 Mean values for 10000 simulations of every combination of random variables simulating indicator ratings, $\mu = 3$ for every Gaussian distribution and Brownian motion with 25 indicator ratings for every parameter

x-axis	y-axis	$\overline{MSD}$	$\overline{SDD}$	$\overline{tt}$	$\overline{td}$	$\overline{ipd}$
Gaussian, $\sigma = 0.25$	Gaussian, $\sigma = 0.25$	0.161	0.149	0.004	0.173	0.814
Brownian, $\sigma = 0.2$	Brownian, $\sigma = 0.2$	0.094	0.152	0.01	0.024	0.229
Gaussian, $\sigma = 0.5$	Gaussian, $\sigma = 0.5$	0.539	0.112	0.025	0.921	0.47
Brownian, $\sigma = 0.5$	Brownian, $\sigma = 0.5$	0.761	0.272	0.032	0.519	0.295
Gaussian, $\sigma = 0.5$	Gaussian, $\sigma = 1$	0.692	0.303	0.034	1.256	0.332
Brownian, $\sigma = 1$	Brownian, $\sigma = 1$	1.078	0.269	0.037	1.118	0.315
discrete uniform	discrete uniform	1.083	0.119	0.046	2.013	0.241
Gaussian, $\sigma = 0.25$	discrete uniform	0.621	0.92	0.032	1.29	0.381

This results in the mean values of the parameters of the ASSG method in Tab. 8.2. Here the combinations for the boundaries of the relation types are displayed, which where covered in Tab. 8.1.

As already mentioned, the results of the χ^2 tests are not finite. For example, for a stationary relation with a fixed y-axis, the result of the χ_x^2 test is infinite (see Fig. 7.1). The strong law of large numbers is therefore not applicable for all χ^2 values. However, in the case of the critical values of the χ^2 tests, the low results of this parameter need to be investigated. Therefore, only the finite results of the χ^2 are considered in the simulation. Therefore, only finite results are considered in the calculation of the mean values, which allows the application of the strong law of large numbers. This leads to the mean values of Tab. 8.3.

Table 8.3 Mean values of χ^2 tests for 10000 simulations for the combination of Gaussian distributed indicator ratings with $\sigma = 0.25$ and uniform distributed indicator ratings with 25 indicator ratings for every parameter

x-axis	y-axis	$\overline{\chi_x^2}$	$\overline{\chi_y^2}$	$\overline{\chi_{lin}^2}$
Gaussian, $\sigma = 0.25$	discrete uniform	7900.434	2.16	7892.943

As expected, the result of the mean value for χ_y^2 is low at $\overline{\chi_y^2} = 2.16$. The other two mean values are very high because the results for the other two χ^2 values can still be finitely high. However, they are not critical values for the χ^2 tests and only represent possible results of the ASSG method. They also do not represent an expected value of the possible outcomes, as there is still the error of not considering the outcome of infinity. In the case of the χ^2_{lin} and the χ^2_x tests, infinity is a likely outcome (see 7.1) that is not included in these mean values of the simulation. Therefore, critical values for high results of the χ^2 tests can not be simulated with the strong law of large numbers. In this case, they are also not needed for the critical values of the flow chart. The electronic supplementary material lists further combinations of random variables for standard deviations between the combinations from Tab. 8.2 in chapter C of the electronic supplementary material.

In a table like Tab. 8.2, a change in the parameters due to a higher number of indicator ratings can not be taken into account. To investigate such behaviour in the parameters, each mean of the parameters would have to be simulated for different numbers of indicator ratings. In addition, this simulation would then have to be carried out for at least the combinations of the random variables of Tab. 8.1.

This leads to plots like Fig. 8.27. Here each point represents a result for the mean of 10000 simulations for one parameter. The value of the x-axis is the number of indicator ratings used for the calculation of the mean value. This represents the length of a video used for the ASSG method and the number of points inside the ASSG method for this video length. The value of the y-axis represents the outcome of the calculated mean value. The colour of each dot then represents the type of parameter that was calculated. For example, the green points represent the outcome of the parameter ipd.

In the last chapter it was shown that indicator ratings below ten ratings for an ASSG plot are insufficient for the application of the weak law of large numbers. These points with a lower number of indicator ratings than ten are marked with an x-symbol. They are also not included in the estimation of a chance for the parameter results as the number of indicator ratings increases. This is because the ASSG method is not recommended for these types of ratings.

To examine the trend of the parameters with increasing number of ratings, a linear function is fitted to the results of each parameter. The results with a lower number of ratings than ten are not included in the linear fit. To assess how good the fit is, a χ^2_{lin} test is performed for all fits. This type of χ^2 is the same as in example 6.7.3.

To what extent the parameters change in this simulation can then be evaluated using the parameter m of the linear fit

$$f(x) = mx + b.$$

To determine the parameter b in this fit, the results of the parameters with 25 ratings can be used, which were already simulated in Tab. 8.3. This leads to this simple calculation:

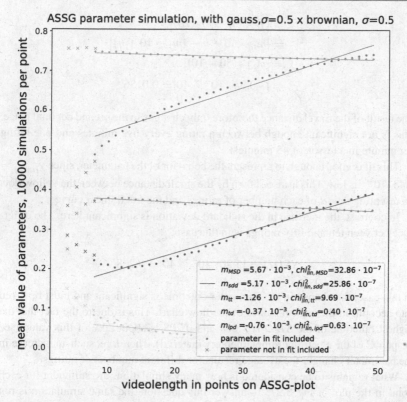

Figure 8.27 Simulation of parameters for one Brownian motion and one Gaussian distribution with $\sigma = 0.5$, 25 ratings per ASSG plot, and 10000 simulations of each parameter per mean value for each point

$$f(25) = m \cdot 25 + b \Leftrightarrow b = f(25) - m \cdot 25. \tag{8.1}$$

Therefore, in the simulations as in 8.27, only the parameter m for the linear fit will be calculated.

If the parameter does not increase or decrease enough to be considered in the flow chart, the result of m will be small. An example of this would be the travel distance in Fig. 8.27. Here $m_{td} = -0.34 \cdot 10^{-3}$. This means that the difference Δtd between an ASSG plot with ten indicator ratings and fifty indicator ratings for the travel distance is:

$$\Delta td = \left| f_{td}(50) - f_{td}(10) \right|$$
$$= \left| m_{td} \cdot 50 + b - \left(m_{td} \cdot 10 + b \right) \right|$$
$$= \left| m_{td} \cdot (50 - 10) \right|$$
$$= \left| -0.34 \cdot 10^{-3} \cdot 40 \right| = 0.0136.$$

The result of the travel distance therefore only changes in the second decimal place. This is not significant enough between a rating every five minutes and one rating per minute in a lesson of 45 minutes.

This fit is good enough to represent the behavior of the parameter, since $\chi^2_{lin,td} = 0.48 \cdot 10^{-7}$ is low. This is also shown by the small distance between the yellow dots representing the td of each number of ratings and the fitted linear curve.

In contrast, the increase in the standard deviation is significant here. The difference between ten and fifty ratings is in this case:

$$\Delta SDD = \left| 5.20 \cdot 10^{-3} \cdot 40 \right| = 0.208.$$

In this case, the change in the parameter outcome is significant and must be taken into account in the critical values of the flow chart. This is due to the fact that the highest outcome of the SDD is 1.084 and 0.208 is about 20% of this value (see chapter C of the electronic supplementary material). Therefore, such an increase in the parameter changes the average results for different numbers of ratings.

What remains to be considered is how many simulations are sufficient for each point in the plot of Fig. 8.27. To answer this question, the same simulation is run for a different number of simulations per mean for each point in the plot. For the combination of two Brownian motions with a standard deviation of $\sigma = 0.5$, 10 simulations (Fig. 8.28), 100 simulations (Fig. 8.29), 1000 simulations (Fig. 8.30), and 10000 simulations (Fig. 8.27) were performed.

In the simulation with only 10 results for each point, it can be seen that the points for each parameter do not linearly converge. This can also be seen in the higher results of the χ^2_{lin} tests. For example, the χ^2_{lin} value for the parameter ipd is more than 300 times higher than the χ^2_{lin} test with 10000 simulations per point.

In Fig. 8.29 with 100 simulations, the distortion of the points becomes smaller and they start to fit better to the linear functions. However, the results for the slope of the linear functions still differ in the third decimal place. For example, the slope of the MSD parameters is almost twice as high compared to Fig. 8.27.

Only in 1000 simulations do the deviations of the parameters seem to match their counterpart in 10000 simulations. Here the largest difference of the slopes is in the

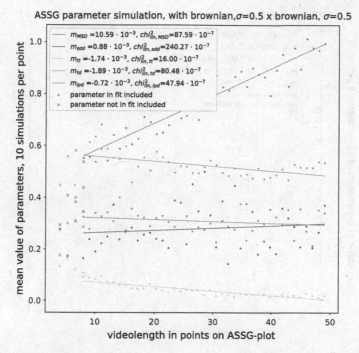

Figure 8.28 Simulation of parameters for one Brownian motion and one Gaussian distribution with $\sigma = 0.5$, 25 ratings per ASSG plot, and 10 simulations of each parameter per mean value for each point

MSD with a difference of $0.05 \cdot 10^{-3}$. Thus 1000 simulations seem to be enough for an estimation of the changes in the parameters.

The tendency of changes in the parameters can also already be seen with 1000 simulations. This is different in Fig. 8.29, where e.g. the results of the td parameters are lower than in Fig. 8.27. 100 simulations are therefore not sufficient to estimate the increase or decrease of the parameters. Therefore, to investigate the trend of the parameters, 1000 simulations are adequate. However, if the results of the slopes are important for the flow chart, a figure with 10000 simulations per point is required. This also means that a simulation with a fixed number of evaluators with 10000 simulations is sufficient, which is why 10000 simulations were chosen for Tab. 8.2.

Therefore, the combinations of Tab. 8.2 will be simulated with 1000 simulations per point and can be found in chapter D in the electronic supplementary material. These figures can be categorised into the two following types: those where there

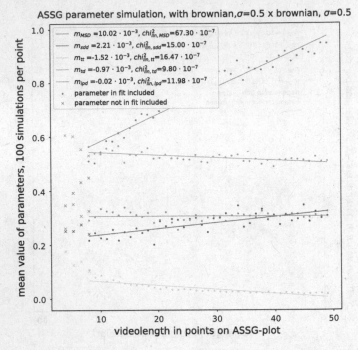

Figure 8.29 Simulation of parameters for one Brownian motion and one Gaussian distribution with $\sigma = 0.5$, 25 ratings per ASSG plot, and 100 simulations of each parameter per mean value for each point

is no significant change in the slope of the linear functions and those where the parameters change significantly enough to be included in the flow chart. In cases where the standard deviations are very high, there is no change in the parameter results. For example, a simulation with a Gaussian distributed rating with $\sigma = 2$ and uniformly distributed ratings does not have a significant enough slope for any parameter (see chapter D of the electronic supplementary material). The other type of simulation is where the slope of the linear function is large enough to be considered, as in Fig. 8.27. This type of simulation will be investigated further while applying the critical values to the flow chart in Fig. 7.10 and Fig. 7.10.

The final version with all critical values of Fig. 8.31 and Fig. 8.32 will be discussed one by one in this chapter.

The first entry in the flow chart is the travel distance. In this step, the ASSG method should result in a "no relation" box if the travel distance is too high for at

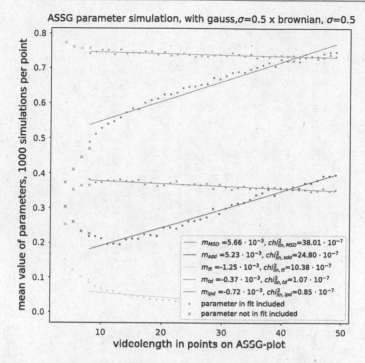

Figure 8.30 Simulation of parameters for one Brownian motion and one Gaussian distribution with $\sigma = 0.5$, 25 ratings per ASSG plot, and 1000 simulations of each parameter per mean value for each point

least a weak relation type. The travel distance therefore has to be higher than for the combinations that are still considered weak relations. These combinations can be found in Tab. 8.1. The travel distance has to therefore be higher than 1.256. Furthermore, there is no significant increase in travel distance in any other simulations (see chapter D of the electronic supplementary material). Therefore, an increase in this parameter for a different number of ratings does not need to be included in the flow chart. However, even such insignificant changes can change the result of the second decimal place of the parameter, as in the calculation for Fig. 8.27. Therefore, the critical values for the flow chart can only be used for the first decimal place. In the case of the travel distance, this means that the critical value needs to be estimated upwards. Which means that ASSG plots with a higher travel distance than for weak relations are still sorted out in this step of the flow chart. The types of ASSG plots that slip through this estimation upwards can still be sorted out in the next steps.

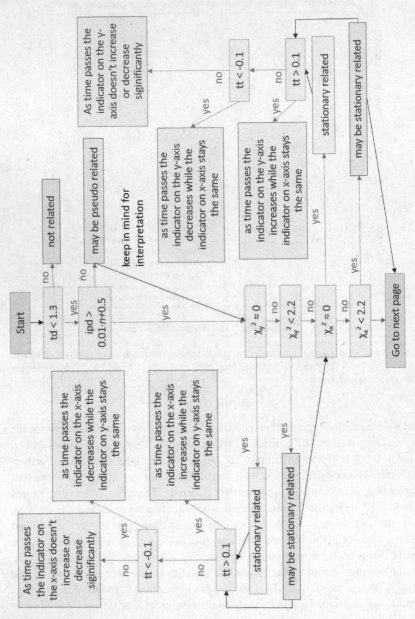

Figure 8.31 Flowchart for interpretation of ASSG-Plots with 5x5 indicators

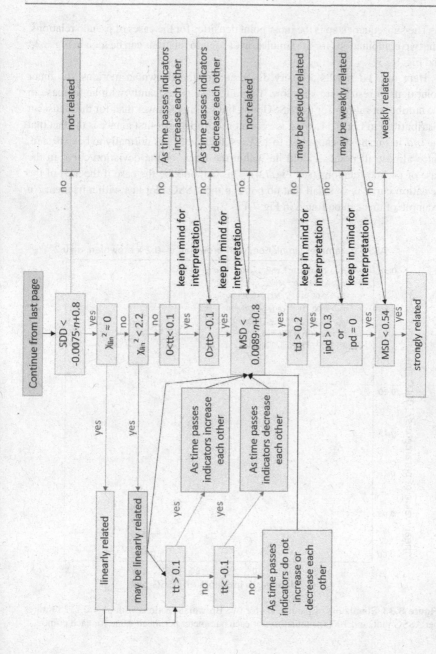

Figure 8.32 Flowchart for interpretation of ASSG-Plots with 5x5 indicators

The subsequent step is the inner point densities for the cases of pseudo relations. The two combinations for the simulation of pseudo relations can be seen in Fig. 8.34 and Fig. 8.33.

Here the ipd results are very different. For the Brownian motions, the inner point density results are very low. They increase significantly with an increase in the number of ratings for the ASSG plot, but are much lower than for the Gaussian distributions in Fig. 8.34. What is even more surprising at first glance is the fact that the ipd in Fig. 8.33 is not close to 1. A result close to 1 is normally to be expected, since almost all points should lie within the range of the deviation curve in the case of pseudo relations (see Fig. 7.4). This can only be the case if the area of the deviation curve is so small that no point in the ASSG plot lies within its area. An example of this can be found in Fig. 7.12.

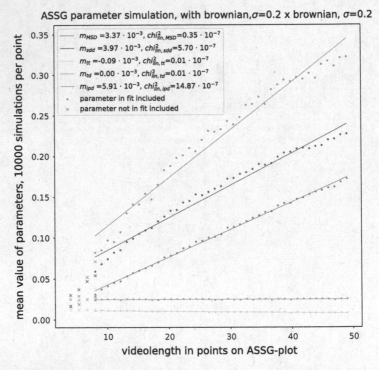

Figure 8.33 Simulation of parameters for two Brownian motions with $\sigma = 0.2$, 25 ratings per ASSG plot, and 10000 simulations of each parameter per mean value for each point

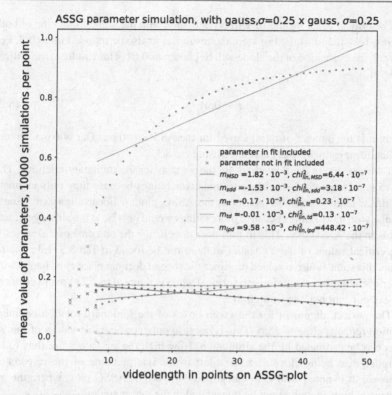

Figure 8.34 Simulation of parameters for two Gaussian distributions with $\sigma = 0.25$, 25 ratings per ASSG plot, and 10000 simulations of each parameter per mean value for each point

Therefore, the estimation of a critical value for the ipd has to be done using Fig. 8.34. The problem with this figure is that this is the only case of all combinations where the parameter results are not a good enough fit to a linear function even with 10000 simulations. This is also evident from the high χ^2_{lin} value of this fit. Furthermore, the case of an ipd of zero can also occur in this case. Therefore, the critical value of the ipd needs to be estimated upwards. The errors made in this estimation can be taken into account in the second categorisation for the pseudo relation. There, the travel distance will give a second insight into the possibility of a pseudo relation. In contrast to the ipd the td does not change with an increase in indicator ratings.

The critical value for the td will be estimated with the highest outcome of both figures. Like the td of the last step, the result has to also be rounded up to 0.2. For the ipd, the estimation of the slope will be chosen as 0.01. This results in the critical value of

$$ipd < 0.01 \cdot n + 0.5, \tag{8.2}$$

where n is the number of ratings used for the ASSG method. The y-axis intercept of this linear estimate was calculated using eq. 8.1.

The next step in the flow chart are the χ^2 tests for the stationary relations. The χ^2 values do not change with more indicator ratings because they only estimate the linear alignment of the points in the ASSG plot. A linear alignment cannot change with more ratings because more ratings would only be at the already aligned points in the ASSG plot. This in turn does not change the outcome of the χ^2 tests. The critical values of the χ^2 tests can therefore be found in Tab. 8.3. Like the td value, this value is also rounded up, since the second decimal place can change with several simulations. Therefore, the critical value is rounded to 2.2 to avoid errors in the categorisation of the flow chart.

The perfect alignment for the green boxes of the stationary relation or linear relation remains close to zero. This is because only considerable values of the χ^2 tests can be estimated by the simulation. How high the χ^2-values are thus to be considered as being close to zero is left to the interpretation of the researcher. However, it is now possible to categorise the types of ASSG plots where the χ^2 values are high enough to not be considered in the interpretation.

The next critical value would be the travel tendency. The tt should not be high enough in any combination of the simulations to be included in the flow chart. This is because no indicator value in this simulation favours dynamic development in any direction. This also means that should the travel tendency be higher than in these simulations without dynamic progression, this must be taken into account in the ASSG method. Therefore, the critical value for the flow chart must be higher than in any result of the simulations.

The absolute slope of the linear fits of the tt in the simulations is highest in Fig. 8.35. Furthermore, there is no simulation with a significantly positive slope of this linear fit (see chapter D). The highest value of the tt is therefore found in Fig. 8.35 with the lowest number of ratings still considered, which is 10. After rounding to the first decimal place the critical value for the travel tendency is 0.1. Since the tt can also be negative for a dynamic progression towards lower values of ratings, the critical value for this case is -0.1. The lowest and the highest still considered values for the travel tendency are 0.1 and -0.1.

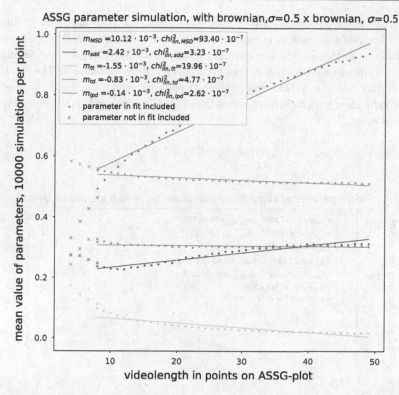

Figure 8.35 Simulation of parameters for two Brownian motions with $\sigma = 0.5$, 25 ratings per ASSG plot, and 10000 simulations of each parameter per mean value for each point

The critical values for the χ_x^2 and the tt can also be applied to the other stationary relation and the linear relation. This is because only the orientation of the alignment changes and not the alignment itself. Therefore, the critical values for the other two cases do not change and can also be used in the flow chart.

The next step in the flow chart is the category where the standard deviation is too high for the other relation types. This case can be simulated with a higher standard deviation than in the case that simulates the stationary relation in combination with uniformly distributed indicator ratings. These would be the Gaussian distributed ratings with a standard deviation of $\sigma = 0.5$ or the Brownian motion with $\sigma = 0.5$. The combination with the Gaussian distribution remains the same over a varying number of ratings (see Chapter D.10 of the electronic supplementary material). When combined with the Brownian motion, the SDD decreases. In this case, downward estimation is required, as any case where the SDD is too high needs to be sorted out. The non-changing results of the SDD in the figure for the Gaussian distribution are higher than the points for Brownian motion. Therefore, the critical value for SDD being too high can be estimated using the linear fit of the combination with Brownian motion from Fig. 8.36. This leads to the following critical value:

$$SDD < -0.0075 \cdot n + 0.8. \tag{8.3}$$

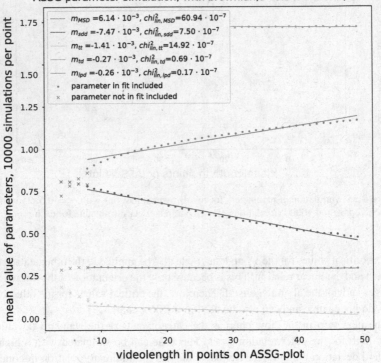

Figure 8.36 Simulation of parameters for one Brownian motion with $\sigma = 0.5$ and one discrete uniform distribution, 25 ratings per ASSG plot, and 10000 simulations of each parameter per mean value for each point

This leaves the last categorisation for no relation with the mean standard deviation. Here, the combinations for the limits to the weak relations are to serve as critical cases. These can be found in Tab. 8.1. Like the SDD, this parameter also needs to be estimated downwards. The MSD remains the same with the Gaussian distributed indicators with an increasing number of ratings (see Fig. D.10 of the electronic supplementary material). The parameter changes in the combination with the Brownian motions (see Fig. 8.37). As with the SDD here the combination with the Gaussian distributed indicator is always higher than with the Brownian motions. Therefore, the linear fit in Fig. 8.37 can be used for the critical value. Similar to fig 8.34, this

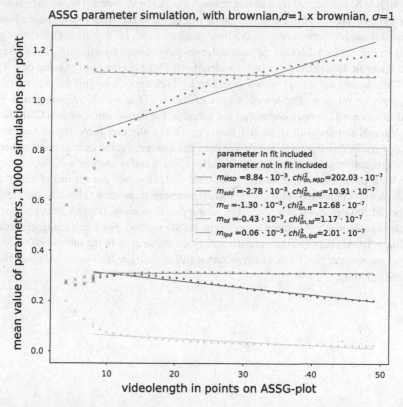

Figure 8.37 Simulation of parameters for two Brownian motions with $\sigma = 1$, 25 ratings per ASSG plot, and 10000 simulations of each parameter per mean value for each point

fit has a higher χ^2_{lin} value. To compensate for this, the y-axis intercept has to be rounded up to the first decimal place. This leads to the following term:

$$MSD < 0.0089 \cdot n + 0.8. \tag{8.4}$$

The last categorisation is the difference between a strong relation and a weak relation. This is where the critical values for the ipd and the MSD have to be found. The boundary between a weak relation and a strong relation was set for combinations of standard deviations of $\sigma = 0.5$ (see Tab. 8.1).

In the case of the ipd, a critical value is needed for which all ipds that are high enough to be considered for a strong relation have to be included. The lowest values for this case are found in the combination of Fig. 8.27. The other two combinations with standard deviations of $\sigma = 0.5$ have results for the ipd that are higher than the case in Fig. 8.27. In this case the parameter does not change significantly. Therefore, the linear fit does not need to be considered. This leads to the critical value of 0.3.

This leaves the MSD which needs to be high enough to still be considered as a strong relation. The lowest values of the MSD at which a combination is still considered strongly correlated are found in the combination of two Gaussian distributed ratings with $\sigma = 0.5$ (see Fig. D.9 of the electronic supplementary material). In this this case, the MSD does not change significantly enough either. The critical value can therefore be found in Tab. 8.2 and is rounded to 0.54.

With the critical value of the MSD, all critical values are estimated and the flowchart is completed. This also answers the research question "How can different types of SSG plots be categorized?". Using this flow chart, the researcher is now able to categorise the relation types of the ASSG method. For a first categorisation of the relations between the indicators, only the parameters of the individual ASSG plots are needed. Any further interpretation still depends on the researcher using the ASSG method.

Comparison between Advanced State Space Grids and a Standard Research Method

9

The aim of this chapter is to answer the third and fourth research questions. These are: "Are there differences in results between the ASSG method and a standard method for indicator based analysis?" and "What are advantages and disadvantages of the ASSG method in comparison to a standard method?".

The standard method used for this comparison with the ASSG method will be the method explained in Section 3.2. The indicators for both methods will be the same and are discussed in Section 9.2. The comparison of the results is then analysed in Section 9.4, which is the first attempt to answer the third research question. A further investigation of this research question will be done in Chapter 10. The following Section 9.5 deals with the advantages and disadvantages of the ASSG method. This serves to answer the fourth research question.

9.1 Data Collection

The type of data collection and the indicators for this chapter were chosen in order to conduct a feasibility study of the ASSG method in comparison to the standard method from 3.2. The indicators for this study were selected from aspects of the basic dimensions of good teaching covered in Chapter 2. The interpretation of the content of this study on the relationships between the indicators and the results of the standard method is limited to the applicability of the ASSG method and its comparison. A more extensive interpretation of the results is not undertaken, as it would not contribute to answering the research questions.

© The Author(s), under exclusive license to Springer Fachmedien Wiesbaden GmbH, part of Springer Nature 2023
N. Litzenberger, *Introduction of Advanced State Space Grids and Their Application to the Analysis of Physics Teaching*, BestMasters,
https://doi.org/10.1007/978-3-658-42732-0_9

The data collection is carried out by four students. Two of them used the ASSG method while the other two use the standard method. Both teams go through a rater training before they start to rate the indicators for their method.

Two different lesson videos are used for both methods. One is a lesson in German that lasts 33 minutes. The other is an English lesson that lasts 49 minutes. For indicators that count specific events during the lesson, this difference in duration must be taken into account. In the ASSG method, this difference only changes the number of ratings that can be made during the lessons. Each indicator is rated every two minutes. For the shorter lesson, this results in 17 ratings for an ASSG plot. This number of ratings is still usable for the ASSG method, but it is not ideal. The 49 minute long English lesson yields 25 ratings. This number of ratings is more suitable for the ASSG method.

For global measurements of the indicators carried out with the standard method, this difference in video length must also be taken into account when ranking the indicators. For example, an indicator that counts the number of times the teacher changes position may give higher results if the measurement duration is longer. This is because the teacher has more time to change his position more frequently. However, the frequency with which he changes his position is different if the number of position changes is the same in both videos. This difference in video length must therefore be taken into account when evaluating the indicators.

To account for this difference, the number of position changes required for a given rating is adjusted. For the adjustment, a formula is used that takes into account how long the total duration T of the video is. In addition, the formula takes into account how often the event must occur at least in order to achieve the comparable rating in the ASSG method in the two-minute interval. This results in the formula for the times N that the event must occur in order to still be considered for the rating of the indicator:

$$N = n \cdot \frac{T}{2}. \tag{9.1}$$

The conversation factor $\frac{T}{2}$ rescales the required events n for an ASSG method rating to the required events N for the standard method. The factor increases if the duration of the measured video T is higher. This leads to the conversion factor

$$\frac{T}{2} = \frac{49}{2} = 24.5 \tag{9.2}$$

for the English lesson. And to the conversation factor

$$\frac{T}{2} = \frac{33}{2} = 16.5 \tag{9.3}$$

for the German lesson.

With this formula, the indicators only have to be defined for the two-minute rating intervals of the ASSG method. The number of events that occurred for a global measurement of the standard method can be calculated with the formula.

9.2 Indicators for the Study

The indicators of this study are divided into the different basic dimensions and their sub-dimensions. In this study, the indicators used are derived from the basic dimensions of *classroom management* and *student support*. If there is a relationship between the aspects of these basic dimensions, then the ASSG method should lead to a relationship type for the indicators representing these aspects.

In addition, the types of indicator rating systems are varied to investigate whether and how different types of indicators can be used for the ASSG method. These types of indicator ratings are normally used for the standard method. Therefore, if the ASSG method has difficulties in using these indicators, this is a disadvantage of the new method.

9.2.1 Basis Dimension: Classroom Management

With the help of classroom management, undisturbed learning, and thus active engagement with the learning material can be created. According to Kounin (2006), one of the preventive measures is ubiquity. This describes classroom behaviour that conveys to the learners that the teacher is aware of the behaviour and actions of the entire class and can intervene if necessary (Lipowsky & Bleck 2019).

The characteristic captures the extent to which the teacher is present in the classroom or "omnipresent" in the sense of Kounin (2006). This includes the teacher's ability to do several things at once (multitasking), whether through instructions, comments or even just glances. The teacher can attend to multiple behaviour problems without losing track or control of the situation. The teacher literally has "eyes in the back of her head" (Seidel 2009). A teacher's omnipresence is particularly evident during the phases when the teacher communicates with individual students.

In this case, it is important to make sure that the teacher has the rest of the class in view (Gabriel 2014).

Ubiquity refers to an omnipresent teacher who conveys presence to learners through their verbal and nonverbal communication. The following indicator includes the area of nonverbal communication as the teacher, using physical presence conveys closeness to learners. Ubiquity can also be expressed by occasional movements through the classroom, e.g. during individual work phases (Gabriel 2014).

In the context of classroom management, the first indicator is primarily intended to measure changing positions, as the teacher adopts different perspectives on the learners. Lotz and Lipowsky (2015) also emphasise in their commentary on the Hattie study that the omnipresence of the teacher can be ensured in the context of classroom management through a dynamic change of position. The change of position can not only have a preventive effect, but can also be actively used in the event of a disruption by the teacher going behind the disruptor. It is furthermore important to check how long the teacher remains standing until the teacher moves on to the next change of position.

Therefore, the first indicator is formulated as follows:

Indicator 1
The teacher moves through the classroom (low inferent)

Rating system

- A 5 is awarded if the teacher completes four or more position changes.
    global rating for English lesson: 98 times or more
    global rating for German lesson: 66 times or more
- A 4 is awarded if the teacher completes three position changes.
    global rating for English lesson: 73 times or more
    global rating for German lesson: 49 times or more
- A 3 is awarded if the teacher completes two position changes.
    global rating for English lesson: 49 times or more
    global rating for German lesson: 33 times or more
- A 2 is awarded if the teacher completes one position change.
    global rating for English lesson: 24 times or more
    global rating for German lesson: 16 times or more
- A 1 is awarded if the teacher completes no position change.
    global rating for English lesson: less than 24 times
    global rating for German lesson: less than 16 times

In order to be able to investigate ubiquity more closely and to check whether different properties of ubiquity are mutually dependent, a further indicator for a second aspect is formulated for this purpose.

Bohl and Kucharz (2010) discuss five possibilities of class room management in open teaching based on Kounin's characteristics. The omnipresence of a teacher in open teaching phases becomes visible, for example, when the teacher advises a group while observing another restless group, or when one has the impression that the teacher is specifically or systematically observing the students and/or groups of students (Bohl & Kucharz 2010, Gabriel 2014). Thus, eye contact functions as a non-verbal means of communication that is of fundamental importance. In practice, eye contact emanating from the teacher can convey a lot of information, such as praising, reprimanding, demanding attention, prompting learners or withdrawing the word. Eye contact can send and receive at the same time; returning the learners' gaze signals to the teacher that they have understood and are ready for the lesson. Thus, eye contact primarily serves to regulate the process of teaching and to cultivate relationships (Döbble 2003). With this non-verbal contact, ubiquity can be realised as a preventive measure in the classroom, as the learners gain the impression that they are being taught by a teacher who keeps an eye on the entire teaching process (Seidel 2009). This happens during the entire teaching process and should be implemented in all phases of work. When collecting and/or handing out work materials, the teacher lets the eyes wander back and forth in the class (Gabriel 2014). As a teacher, it is therefore important to use eye contact appropriately in order to subtly and continuously convey omnipresence. The teacher mainly focuses her gaze on her students and does not turn her back on them too often. For example, even when writing on the board, the teacher repeatedly turns her gaze to the class (ibid.).

Therefore, the second indicator is chosen as follows:

Indicator 2
The teacher's gaze is mainly directed towards the students. (low inferent).

Rating system

- A 5 is awarded if the teacher looks around the class almost exclusively, looks at different students, and keeps looking at the class when writing on the board.
- A 4 is awarded if the teacher frequently looks around the class, looks at different students, and keeps looking at the class when writing on the board.

- A 3 is awarded if the teacher occasionally lets the gaze wander around the class, looks at a few students, and occasionally turns his or her gaze back to the class when writing on the board.
- A 2 is awarded if the teacher rarely looks around the class, looks at few students, and rarely looks at the class when writing on the board.
- A 1 is awarded if the teacher does not look around the class at all or looks almost exclusively at specific students and never looks at the class when writing on the board.

Kounin already took up transition management in the context of classroom management and examined smoothness and dynamics as a control of teaching processes (ibid.). This characteristic is closely related to the characteristic productivity. A high rating of the characteristic productivity implies that transitions are routine and little time is wasted. In the literature, smooth and structured transitions are characterised by the teacher completing the students' previous work, making sure that the students put away their materials, and observing the students' social behaviour during the phase and intervening if necessary (ibid.).

This also means that students always know what tasks are set at that exact moment and can start directly. These smooth and structured transitions often work much better in older classes, as younger students usually need more time and practice (routines and rituals). In terms of general lesson planning, smooth and sweeping classroom transitions not only have an impact on the quality of teaching in general, but also on students' learning motivation and learning efficiency (Kounin 2006, Lipowsky & Bleck 2019). Especially the dimension of smoothness and dynamism requires a good preparation of the lesson by the teacher so that central complications such as hecticness or boredom can be avoided (Gabriel 2014).

In Evertson et al. (2006), the structuring of lessons is addressed with regard to optimal classroom management. Here, the aspect of optimal and effective use of time is considered an essential aspect of effective classroom management and is determined to be a central predictor of learning success (Seidel 2009).

Thus, the better the teacher succeeds in using the available time for subject-related work and learning and in outsourcing so-called "non-instructional activities" or time wasting activities from the lesson, the more active learning time is available (Gabriel 2014). In order to structure a lesson well, in addition to time management, a clear sequence and preparation for possible difficulties must be ensured. Evertson et al. (2006) refer not only to the preparation of the classroom, but also to the accessibility of materials as well as the use of aids.

Consequently, this indicator is formulated as follows:

Indicator 3
The teacher structures the lesson section. (low inferent).

Rating system

- A 5 is awarded if the teacher visualises the structure of the lesson section and provides a time statement.
- A 4 is given if the teacher discusses the structure of the lesson section.
- A 3 is given if the teacher makes the structure of the lesson section clear using a time statement.
- A 2 will be given if the teacher identifies lesson section phases.
- A 1 will be given if the teacher does not identify lesson section phases.

In the following, the aforementioned rules and rituals in the classroom and their added value in the school context will be examined in more detail. Rituals are often a symbolic act, but they promote the development of security through routine and can thus contribute to the promotion of learning (Kippert 2008). The opposite is the case when rules are used. A rule is understood to be a regulation (e.g. students don't just get up and go to the toilet) that implies a clear expectation of behaviour. Wannack (2012) states that rules are used to maintain the integrity of the individual student, communication in the classroom and between children, care in handling materials, orderliness of personal belongings, and mobility inside and outside the classroom.

The consideration of rules and rituals is not only of great importance in the school context. Establishing and adhering to behaviourally effective rules is considered a central factor of teaching and learning, as this not only serves to avoid disruptions, but also to activate the attention and motivation of learners (Helmke 2006, Gabriel 2014).

In order for a ranking system to be established, certain properties of rules and rituals are taken into account for this purpose and it is counted how many of these properties are fulfilled. For this purpose, the following properties are taken into account:

1. students sit in places and do not stand up without being asked,
2. students do not talk when someone else is talking,
3. students use conversational methods (reporting and being called on, talking quietly among themselves in a group work phase, or concrete methods that have been discussed beforehand),

4. students are friendly with each other (use friendly conversational tone when talking, or do not insult each other).

Using these properties, the following indicator is formed:

Indicator 4
Students follow rules and rituals. (high inferent).

Rating system

- A 5 is awarded when all four characteristics are met.
- A 4 is awarded when three characteristics are met.
- A 3 is awarded when two characteristics are met.
- A 2 is awarded when one characteristic is met.
- A 1 is awarded when no characteristic is met.

9.2.2 Basis dimension: student support

The support of the teaching climate is a multi-layered construct with different conceptualisations and operationalisations (Lipowsky & Bleck 2019). Eder (2002) also emphasises that the metaphor of climate to describe the quality of social conditions in the school context is difficult to delineate and accordingly has a great diversity. Not only the subjective learning environment is addressed here, but also the school culture and the school ethos from a pedagogical point of view. With regard to the teaching climate, the quality of the relationships of the actors involved as well as the quality of the professional-adaptive support of the learners in the learning process is considered (Adelman & Taylor 2005, Gruehn 2000). Furthermore, in a good teaching climate, the teacher signals caring, interest in the students' concerns, and creates a good relationship of trust and interaction with empathy.

Eder (2002) emphasises the importance of differentiating between school and class climate and cites the following aspects as supporting elements of the class climate concept:

- The physical environment of the class, equipment, and quality of the facility.
- The social relationship between teacher and learners and the social relationship among learners.
- The expectation regarding performance and behaviour.

- The way teaching and learning processes are carried out.
- The specific norms and values that apply in the classroom.

According to this, in the course of a good classroom climate, students experience a sense of autonomy, competence, and social inclusion, which promotes self-determination in class and has a positive effect on their engagement and motivation in class (Lipowsky & Bleck 2019). According to previous studies, constructive learning support (including effective classroom management) strongly depends on the pedagogical-psychological knowledge of the teacher (ibid.). In addition to the Austrian researchers König and Pflanzl (2016), Neuber, Gebhard and Lipowsky (2015) have also found that a teacher with a higher level of pedagogical-psychological knowledge and more favourable prerequisites in terms of personality traits and motivational characteristics is more likely to be able to create constructive learning support and a positive classroom climate.

A central part of communication in the classroom is the expression of feedback. According to the Hattie study, feedback is one of the ten most important characteristics of learning success and thus one of the most effective factors for learning (Hattie 2009). Furthermore, feedback affects motivation and performance as well as motivational and metacognitive processes (Huth 2004, Lotz, Gabriel & Lipowsky 2013). The distinction between praise and feedback should not be ignored. Praise is usually used as simple affirmation, e.g. "well done" or "good". When specific and differentiated praise is given, it is feedback (Lotz, Gabriel & Lipowsky 2013). Gabriel (2014) clarifies the importance of praise by referring here to the teacher's recognition of the student. Besides attention, warmth, encouragement, factual-constructive feedback, and positive handling of mistakes, the use of praise is a central part of a good teacher-student relationship (Gabriel 2014). Student-student relationships are also characterised by mutual recognition and a positive climate for making mistakes.

In order to measure such positive reactions, the following indicator was chosen:

Indicator 5
The teacher responds with affective, positive reactions to student activities. (low inferent).

Rating system

- A 5 is awarded if the teacher responds more than four times with an affective positive response to a student activity.

global rating for English lesson: 98 times or more
global rating for German lesson: 66 times or more

- A 4 is awarded if the teacher responds more than three times with an affective positive response to a student activity.

global rating for English lesson: 73 times or more
global rating for German lesson: 49 times or more

- A 3 is awarded if the teacher responds more than two times with an affective positive response to a student activity.

global rating for English lesson: 49 times or more
global rating for German lesson: 33 times or more

- A 2 is awarded if the teacher responds more than once with an affective positive response to a student activity.

global rating for English lesson: 24 times or more
global rating for German lesson: 16 times or more

- A 1 is awarded if the teacher does not respond with an affective positive response to a student activity.

global rating for English lesson: less than 24 times
global rating for German lesson: less than 16 times

The teacher's response to classroom contributions is an important aspect of a good student support. In the case of poor contributions, the students should, if possible, come to the correct result on their own through an impulse from the teacher (Kobarg & Seidel 2003). This leads to a sustainable learning process. Criticism is often voiced at this point, as students often only contribute to the flow of the teacher's speech as "cue givers" (ibid.). They are given little opportunity to develop their own ideas (ibid.). A positive attitude towards teaching contributions, including faulty teaching contributions, is a central aspect of cognitive learning support and the student-teacher relationship. Especially mistakes should not be seen as a deficit and blamed, but should be seized as a learning opportunity, which should lead to more effective learning processes (Lipowsky & Bleck 2019, Brophy 2000, Helmke 2012). Furthermore, an appreciative attitude of the teacher towards teaching contributions should trigger cognitive, motivational, and metacognitive processes in the students (Huth 2004, Rakoczy et al. 2008). Ryan and Deci (2000) also confirm that intrinsically motivated student activity increases when the teacher responds appreciatively and with interest to student contributions.

This leads to the sixth indicator which uses Likert scales:

Indicator 6
The teacher responds appreciatively to classroom contributions. (high inferent).

Ranking system

- A 5 is awarded if this is fully applicable.
- A 4 is given if this is more likely to be true.
- A 3 is given if this is partially true.
- A 2 is given if this is more likely not to apply.
- A 1 is assigned if this does not apply at all.

Since this indicator generates greater subjectivity due to this fuzzy separation, additional criteria for the evaluation are defined, which can be found in the electronic supplementary material in Chapter B.1.

9.2.3   Reference Indicator

In addition, a final indicator is formulated, which is formed independently of the relevance and significance for teaching. The aim of this indicator is to show the greatest possible discrepancy with the other indicators. In addition, it should be taken as a reference for the ASSG method for a non-existent relation. For this purpose, the indicator should describe a random interaction as far as possible. Therefore, it is considered how often the teacher touches an object:

Indicator 7
The teacher touches new objects with the hand. (low inferent)

Ranking system

- A 5 is awarded if the number of new objects the teacher touches is over four.
- A 4 is awarded if the number of new objects the teacher touches is four.
- A 3 is awarded if the number of new objects the teacher touches is three.
- A 2 is awarded if the number of new objects the teacher touches is two.
- A 1 is awarded if the number of new objects the teacher touches is one or zero.

Since this indicator is only used for the analysis with the ASSG method, no global rating is required.

9.3   Results for the Relation Types

This chapter will look at the application of the ASSG method to the two rated lessons in more detail. All indicators in the last chapter are rated every two minutes. The ASSG method method is then applied to these ratings.

Each combination of indicators is then categorised using the flow chart in Fig. 8.31 and Fig. 8.32. Which box is fulfilled in the categorisation of each combination can be found in Chapter B.2 of electronic supplementary material. The final result of this categorisation for the English class is shown in B.4 of electronic supplementary material. The combinations with the reference indicator are not included in this table. They will be covered separately.

It is noteworthy that the only two types of relationship found are the weak relation and no relation. This means that there are no pairs of indicators that are strongly related, where the strong relation may be due to a pseudo relation. The weak relations are less likely to be stable due to changes in the proximity of an attractor. This is because the ASSG plots can display sufficiently dynamic progressions.

An ASSG with a strong relation can be found in the categorisation of the German class in Tab. 9.2. However, in this case, no blue box for the pseudo relation was gone through during the categorisation (see Chapter B of the electronic supplementary material). The exact pair with a strong relation in the German lesson also has a weak relation in the English lesson. This reduces the chance that the relation is strong because of the rating itself.

A stationary relation is also found in the German lesson. In this case, the χ_y^2 test results in a value of 0.15, which is far below the critical value of 2.2. In this case, the χ_{lin}^2 test also results in a value of 0.15. This is the case because a stationary relation with a fixed y value is also a linear relation with $m \approx 0$.

Thus, the ASSG method is thus able to successfully categorise the different relation types with the help of the flow chart. There are also only minor differences between the relation types of the two lessons.

Almost all pairs with a weak relation in German classes are also weakly related in English classes. An exception is the pair with the indicators *look through class* and *class phase structured*. This combination is unrelated in the German lesson ratings, but weakly related in the English lesson. There are also five combinations of indicators that are unrelated in the German lesson but related in the English lesson. This can have many different reasons. One reason could be the fact that the ratings were only carried out by two raters. Therefore, the difference in the ratings may be influenced by the subjective views and errors of the two raters (Tab. 9.1, Tab. 9.2, Tab. 9.3).

Another reason could be that the teaching and the behaviour of the two teachers differ so much that there is no influence on the two indicators. Thus, the indicators in the English class do not affect each other, whereas they can influence each other in the German class. However, these differences only exist between two relation types that differ only slightly in their parameters. For example, the MSD is very high at 0.84 when combined with the indicators *glance through class* and *class phase structured*. The difference between no relation and a weak relation is determined with a difference of 0.15, which is not much. This also applies to the other cases.

Moreover, these differences only exist in cases where the indicators do not belong to the same sub-dimension. Each combination with the same sub-dimension is weakly related. For example, the combinations with the *position changes* and *glance through class* are from the same sub-dimension *ubiquity* and are weak related in both lessons. The weak relation is also meaningful in this case. When a teacher moves around the room, it is more likely that the teacher has to at least look around to avoid colliding with a table. The same is true for the two indicators of the basic dimension student support. In both lessons, these two indicators show a weak relation.

An other reason why the ASSG method is suitable for an analysis of the student-teacher system is the fact that almost all combinations with the reference indicator result in no relation (see Chapter B of the electronic supplementary material). In the English lesson, there are only two combinations where the indicator in combination with the reference indicator results in a weak relation. In these two cases, the parameters that distinguish between a weak relation and no relation are very close to the critical value of the flow chart. All other combinations with the reference indicator result in no relation. In particular, all combinations with this indicator are unrelated in the German lesson.

In summary, it is possible with the ASSG method to categorise the indicator combinations into different types of relations. Any combination where a weak relation or a strong relation seem logical is categorised as such. The same applies to almost all combinations with the reference indicator that should not result in a relation. The difference between a weak relation and no relation is different in some cases where the parameters are close to the critical values. In these cases, a researcher using the ASSG method is required to take a closer look at the ASSG plot.

Table 9.1 Categorisation of the relation types, English lesson

relation type	weak relation	no relation
examples	Indicatorset: subdimensions, english class, MSD: 0.78, SDD: 0.39 rating values, td=0.72, tt=0.02 end value start value deviation curve, ipd=0.30 EV: $(3.47, 3.25)^T$ m =0.16, x_m^2=1.32 y =3.22, x_y^2=1.17 x =3.57, x_x^2=10.24 Indicator: glance through class, SD: 0.58 (y-axis) Indicator: position changes, SD: 0.97 (x-axis)	Indicatorset: subdimensions, english class, MSD: 1.24, SDD: 0.38 rating values, td=1.00, tt=-0.02 end value start value deviation curve, ipd=0.22 EV: $(4.01, 2.89)^T$ m =-0.97, x_m^2=3045.70 y =3.00, x_y^2=2899.16 x =3.91, x_x^2=11.16 Indicator: positive reaction, SD: 1.44 (y-axis) Indicator: classphase structured, SD: 1.05 (x-axis)
combination of the indicators	*position changes* x *glance through class*	*position changes* x *class phase structured*
	position changes x *rules and rituals*	*position changes* x *positive reaction*
	position changes x *rules and rituals*	*position changes* x *positive reaction*
	position changes x *appreciation contribution*	*glance through class* x *positive reaction*
	glance through class x *class phase structured*	*class phase structured* x *positive reaction*
	glance through class x *rules and rituals*	*rules and rituals* x *positive reaction*
	glance through class x *appreciation contribution*	*positive reaction* x *appreciation contribution*
	class phase structured x *rules and rituals*	
	class phase structured x *appreciation contribution*	
	rules and rituals x *appreciation contribution*	

Table 9.2 Categorisation of the relation types, German class, part 1

relation type	strong relation	weak relation	stationary relation	no relation
examples	*(chart: Indicator appreciation to learner contribution, SD: 0.47)*	*(chart: Indicator appreciation to learner contribution, SD: 0.47)*	*(chart: Indicator: rules and rituals, SD: 0.46)*	*(chart: Indicator: classphase structured, SD: 1.11)*
combination of the indicators	*rules and rituals* X *appreciation contribution*	*position changes* X *glance through class*	*class phase structured* X *rules and rituals*	*position changes* X *class phase structured*
		position changes X *rules and rituals*		*glance through class* X *class phase structured*
		position changes X *positive reaction*		*class phase structured* X *positive reaction*
		position changes X *appreciation contribution*		

Table 9.3 Categorisation of the relation types, German class, part 2

relation type	strong relation	weak relation	stationary relation	no relation
examples	*(state space grid diagram: Indicator: appreciation to learner contribution, SDI: 0.47; Indicator: rules and rituals, SDI: 0.46)*	*(state space grid diagram: Indicator: appreciation to learner contribution, SDI: 0.47; Indicator: positive reaction, SDI: 0.81)*	*(state space grid diagram: Indicator: rules and rituals, SDI: 0.46; Indicator: classphase structured, SDI: 1.11)*	*(state space grid diagram: Indicator: classphase structured, SDI: 1.11; Indicator: position changes, SDI: 0.9)*
combination of the indicators		glance through class X rules and rituals X positive reaction		
		glance through class X appreciation contribution		
		glance through class X class phase structured X appreciation contribution		
		rules and rituals X positive reaction		
		rules and rituals X appreciation contribution		

9.4 Comparison between the Results

The results of the standard method can be found in Chapter B of the electronic supplementary material. Unlike the ASSG method, no relation between the indicators can be found with the standard method in this case. One possibility for a relation between two indicators that can be found using the standard method would be a correlation study (see Section 3.5). However, far more than two rated lessons would be needed to find a correlation between two indicators. Therefore, the results of the relationships between the indicators cannot be compared between the ASSG method and the standard method.

The results for the global ratings can be compared with the standard method and the results of the expected values of the indicators in the ASSG method. The results for the EV and SD can be found in any of the combinations with the indicators in Chapter B of the electronic supplementary material. The td was calculated separately with eq. 7.1. The results of both methods can be found in Tab. 9.4 and Tab. 9.5. The difference between the two global estimates of the indicators results from the absolute difference between the global rating of the standard method and the expected value of the ASSG method. If both methods lead to a different global estimate of the indicator because the difference is too large, the difference is highlighted in italics (Tab. 9.5).

As can be seen in both tables, half of the results for the global estimation of the indicators differ so much that they lead to different results for the indicators. For only one of the indicators the difference is small in both lessons. This indicator is *glance through the class*. Every other indicator is different enough in at least one of the lessons to lead to different results in both methods.

Table 9.4 Comparison of the results in both methods for the English lesson

indicator	standard method	ASSG			difference
	global rating	EV	SD	td	
position changes	4	3.47	0.97	0.55	*0.53*
glance through the class	3.47	3.25	0.58	0.36	0.22
class phase structured	3.9	4.01	1.05	0.41	0.11
rules and rituals	5	4.48	0.5	0.23	*0.52*
positive reactions	3	2.89	1.44	0.77	0.11
appreciation to contribution	3.26	3.28	0.69	0.27	0.02

Table 9.5 Comparison of the results in both methods for the German lesson

indicator	standard method	ASSG			difference
	global rating	EV	SD	td	
position changes	4	3.11	0.9	0.6	*0.89*
glance through the class	4.01	3.95	0.82	0.4	0.06
class phase structured	3.625	3.21	1.11	0.4	*0.415*
rules and rituals	4.5	4.69	0.46	0.13	0.19
positive reactions	3	3.52	0.81	0.33	*0.52*
appreciation to contribution	3.74	4.34	0.47	0.2	*0.6*

These differences also do not seem to depend on the type of indicator. For example, the indicator *position changes*, which counts a specific event, has high differences in both lessons, whereas the indicator with Likert scales has a difference of 0.02 in the English lesson. It also does not seem to depend on the SD or the td of the indicators. For example, the indicator *glance through the class* has low differences in both lessons, but has almost twice the standard deviation and travel distance than the indicator *appreciation to contribution* in the German lesson.

This can be explained by various reasons. One reason could be the fact that the raters of the standard method and the raters of the ASSG method interpret the indicators differently. For example, it could be that one group interprets a slight movement of the teacher by twenty centimetres as a change in position, while the other group does not consider this a change in position. This could lead to fewer counts for the position changes and ultimately to different ratings. Before both groups started to rate the indicators, a rater workshop was conducted. Nevertheless, such differences can occur when different people rate the same indicator separately.

Another reason could be the fact that only four people evaluated the indicators. Even if they do not interpret the indicators differently, they may make mistakes in their ratings. Such errors can make a bigger difference when only two people evaluate the indicators.

Apart from the errors made by the ratings, there may also be a significant difference between the two methods. For example, it is possible that in the standard method individual events have greater significance than in the ASSG method. Such a case would be, for example, if in the standard method a rating is argued through the use of comparative examples. Such a rating was done for the indicators *rules and rituals* and *positive reactions* (see Chapter B of the electronic supplementary material). Such comparative examples have less influence in the ASSG method, as each comparative example can only influence a single rating in a single time interval. Therefore, the differences in both methods are greater in such cases.

Which of these possibilities is the reason for the differences in the results of the two methods remains unknown for this study. It would be possible to reduce the impact of the first two possibilities if more raters rated the indicators and if the ratings for both indicators were done by the same person. Therefore, another study needs to be conducted to find out the reason for this difference in the results. Such a new study is conducted and analysed in Chapter 10.

9.5 Comparison between the methods

In this chapter the research question "What are advantages and disadvantages of the ASSG method in comparison to a standard method?" is addressed. The difference in results is not considered in this comparison, as this aspect is dealt with in more detail in another study in Chapter 10.

The specific aspects covered in this comparison are the

1. flexibility,
2. preparation needed,
3. the specificness of the implementation,
4. time consummation,
5. and opportunities for knowledge

of both methods.

The requirements for the indicators in the ASSG method have already been discussed in Section 7.5. All these requirements do not exist in the standard method. For example, the standard method does not require each indicator to be rated at least ten times or even more. This means that the standard method is not limited to sufficiently long video recordings. Even a video with a length of one minute can still be rated with the standard method, as only one rating is required for a result. Such a short video would not be suitable for the ASSG method. The display and analysis of a dynamic progression would not be of any use if no dynamics can be displayed with only one rating. This fact makes the ASSG less flexible than the standard method.

The flexibility is also related to the necessary preparation for the ASSG method. Due to the many requirements of the ASSG method, it is not easy to find a useful indicator for a certain aspect of the student-teacher system. For example, an indicator that rates whether a certain aspect is fulfilled or not during the lesson would not be suitable for the ASSG method. Such an indicator can only have two different ratings and can therefore not display a dynamic progression.

On the other hand, the standard method is very flexible in terms of how the indicators are rated, so preparation is needed to decide how the indicators should be rated. For example, it is not concrete how an indicator such as *positive reactions* shall be rated in the standard method. There are various possibilities to decide between. For example, this indicator could be rated separately in each phase of the lesson, just as the indicator *class phase structured* was rated in a structured way. It would also be possible to assess each individual reaction according to how positive it is and then to calculate an average value. Another possibility is to rate the indicator using comparative examples, as was done in this case. However, which choice is best in this case is left to the rater and must be decided in a preparation. In the ASSG method, the type of rating is very specific. The only choice that has to be decided is how long each time interval for a rating should be. This decision is also partly taken away from the researcher, as at least 20 ratings are recommended.

The time required for both methods is approximately the same. The ASSG method needs more time for the selection of indicators. The standard method needs more time for preparation to decide how to rate the indicators. The time needed to rate the indicators is also the same, as the whole video has to be watched to rate each indicator. In this study, at most two similar indicators could be rated at the same time. Since the number of indicators remains the same for both methods, the number of views of the video also remains the same. The standard method took a little more time in this study because more time was needed to filter out the comparative examples. Furthermore a little more time was needed to argue the final decision of each indicator. However, the categorisation of the indicator combinations into the different relation types took additional time that was not required for the standard method. Thus, the time required for both methods is approximately the same and only differs by less than one hour.

A major advantage of the ASSG method is that it offers more possibilities for further insight. The global estimates of the outcome of an indicator can be made with the expected value of the ASSG method. This is also the only measurement that can be made with the standard method. In addition to the global estimate, the ASSG method also provides information on the variability of the indicator's ratings. For example, the indicator *positive reactions* in the English lesson has the highest standard deviation and also the furthest distance. This means that this indicator is very variable during the lesson and also jumps a lot between different ratings. These are findings that cannot be compared between different indicators using the standard method because it does not have comparable numerical parameters.

Another way of knowing is that different types of relations can be categorised by the flow chart of the ASSG method. Since the parameters used for the flow chart are numerical, this first categorisation is also very easy to do. This also means that it does not take much time to categorise all the combinations of indicators. In the case of this study, it only took about 1.5 hours to categorise all 44 combinations.

To find a relationship between two indicators, a correlation study with the standard method would require more rated lessons. With the ASSG method, only one lesson is needed to suspect a relationship between two indicators. Therefore, the amount of data needed to analyse the relationships between indicators is drastically reduced.

All summarised advantages and disadvantages can be found in Tab. 9.6. As can be seen in the table, the ASSG method has many advantages over the standard method. However, the ASSG method is also very limited due to the many requirements for the selection of indicators. Therefore, if a study cannot meet all these requirements, all the advantages of the ASSG method are not applicable. However, if the indicators meet the requirements of the ASSG method, there is no reason not to use the ASSG method.

Table 9.6 Advantages and disadvantages of the ASSG method in comparison to the standard method

Advantages	Disadvantages
+ how the indicators shall be rated is very specific	− less flexible in terms of applicability
+ total time consumed for the method remains the same	− less indicators are usable
+ increase in opportunities for knowledge	− finding a usable indicator is more complicated
+ global estimation also takes the variability into account	
+ parameters can be compared	
+ categorisation in relation types	
+ less data is needed to analyse relationships between indicators	

Differences between Global Indicator Ratings and Global Results of Advanced State Space Grids

In this chapter, the differences from Section 9.4 between the global indicator ratings of the standard method and the global results of the ASSG method are examined in more detail. To further analyse these differences, a second study was conducted. Since only the differences between the two methods are analysed, no further investigation of the relationships between the indicators is carried out. This also applies to the categorisation into different relationship types. Therefore, the relationships of the indicators with the basic dimensions of good teaching are not examined.

10.1 Data Collection

In this study, three different groups of students were asked to rate four indicators. Each group of students evaluates a different video of a lesson. First, the students are asked to watch the video of the lesson and rate the indicators once for the entire duration of the video. This will be the result of the overall rating for the standard method. Then the students will watch the video again and rate the indicators every two minutes for the ASSG method.

The ratings are collected using an online data collection programme called FormR. The interface of this programme can be seen in Fig. 10.1.

In this interface, each rating can be selected by clicking on the corresponding number for each indicator. A box like in Fig. 10.1 is used for each time interval and also for the global rating at the beginning.

N. Litzenberger, *Introduction of Advanced State Space Grids and Their Application to the Analysis of Physics Teaching*, BestMasters, https://doi.org/10.1007/978-3-658-42732-0_10

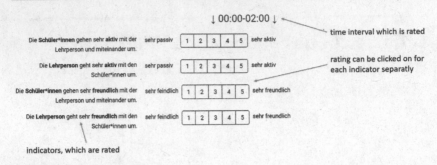

Figure 10.1 Interface of FormR

The students carry out this rating as part of a seminar in the education programme, which is held in German. This is also the reason why the FormR interface and all other documents for the student groups are in German. In this seminar, where the students rate the indicators, the students learn how indicators are formulated and how an analysis of a lesson is carried out with the help of indicators. This part of the seminar is therefore the second time that the students rate the indicators themselves.

After the students have rated all the indicators, they are shown the difference between their overall rating and the ASSG method. In this way, students have the opportunity to see how big the difference is between their first rating after watching the video once and their rating at different time intervals. They can also practise an indicator rating of a whole lesson. Such a rating with their own indicators will also be the task that concludes this seminar. Therefore, this rating serves as a preparation for this task.

In addition, students are asked to complete an evaluation of this part of the seminar and indicate whether they are willing to share their ratings for use in this study. A total of 48 students were willing to share their results, therefore 23 ratings were given for the first lesson, 9 ratings for the second lesson, and 16 ratings for the last lesson.

The evaluation found that the interface for the ratings was clear and easy to use. Most students felt that the course was helpful for their ability to rate indicators. However, many students complained that the rating of indicators took a lot of time as they had to watch the video twice. This could be improved by having the students rate their global values of the indicators after rating the ASSG method. This way they would not have to watch the video twice. However, this way a first impression of the lesson would not be possible for a global rating of the indicators. By the time the students give the global rating, they have already rated the indicators many times in the ASSG method. Thus, the global rating can be influenced by the ratings already made for the ASSG method.

In preparation for this part of the course, students were given a document explaining the use of FormR and the indicators. This document can be found in chapter E of the electronic supplementary material. In addition, they received a separate document which rated a lesson other than their own. This rating was done in more detail to give another example of how the indicator should be rated. These documents can also be found in chapter E.

Two of the videos are the same ones that were used in the last study on Chapter 9. The third video shows a different German lesson of 45 minutes. This results in 23 ratings for the ASSG method, which is sufficient and above the recommended 20 ratings.

10.2 Indicators

The indicators for this study are four very different aspects of the basic dimension *student support*. In this study, two indicators of the *agency* sub-dimension and two other indicators of the *communion* sub-dimension are rated. These two sub-dimensions describe the interaction between the students and the teacher. "Agency suggests that someone is becoming individuated, dominant, has power and control, whereas communion means someone is social, shows love union, friendliness, and affiliation" (Wubbels et al. 2015, p. 366). A high level of agency does not mean that the teacher adopts a controlling or strict behaviour. It is about creating a structured social environment with clear expectations that supports student autonomy. A high level of community gives students recognition and respect, which in return encourages acceptance of rules. Such a learning environment supports the development of student autonomy and engagement (Scherzinger et al. 2020).

Wubbels' group showed that students who perceived greater control and warmth from the teacher showed better cognitive performance, greater engagement, and more positive subject-related attitudes (Wubbels et al. 2015). A first analysis of the *agency* and *communion* with the help of the State Space Grid method was conducted by Scherzinger et al. 2020. They showed that a negative relationship rating from the student perspective is associated with observed interactions that are strongly controlled and directed by teachers, while students themselves tend to be more passive (Scherzinger et al. 2020).

The indicators used by Scherzinger's group are also used in this study. These indicators were chosen because the content aspect of the indicators is not relevant for this study. However, since these indicators were applied in the SSG method, they fulfilled the requirements for the indicators from Section 7.5. Therefore, the applicability of these indicators for the ASSG method has already been tested, which is another reason for their selection. For the two indicators for the *agency* sub-dimension, a distinction is made between the teacher's agency and the students' agency. The indicator for students *agency* is formulated as follows:

Indicator 1
Students are very active/dominant in their interactions with the teacher and with each other.

The *agency* of the teacher will be measured by indicator 2 as follows:

Indicator 2
The teacher is very active/dominant with the students

The ranking system for both indicators was adopted from Scherzinger's study. Since this is a German study and the students evaluating these indicators are also German, Scherzinger's ranking system is not translated. Instead, the ranking system is in German and can be found in Tab. 10.1.

The same categorisation is done for the sub-dimension *communion*. The *communion* of the students will be measured by indicator 3:

Indicator 3
The students are very friendly with the teacher and with each other.

Indicator 4 will be used for the *communion* of the teacher.

Indicator 4
The teacher is very friendly with the students.

As for the *agency*, Scherzinger's ranking system is also adopted for these two indicators (Tab. 10.2).

Table 10.1 Ranking system for the two indicators of the sub-dimension *agency* (Scherzinger et al. 2020)

	Lehrperson	Schülerinnen und Schüler
5	Die Lehrperson verhält sich sehr aktiv, bestimmend und führt stark (z. B. befiehlt, fällt Lernenden ins Wort, schreit)	Die Schülerinnen und Schüler verhalten sich sehr aktiv, bestimmend, versuchen das Unterrichtsgeschehen zu beeinflussen und zeigen viel Eigeninitiative (z. B. machen Vorschläge oder äußern Bemerkungen, widersprechen der Lehrperson)
4	Die Lehrperson verhält sich eher aktiv und kontrollierend und gibt nur sehr wenig Kontrolle über das Geschehen im Klassenzimmer an Schülerinnen und Schüler ab (z. B. interveniert, gibt Anweisungen)	Die Schülerinnen und Schüler verhalten sich aktiv und zeigen Eigeninitiative (z. B. fragen bei Unklarheiten nach, holen sich Unterstützung oder Materialien)
3	Die Lehrperson lenkt zwar, aber sie gibt einen Teil der Kontrolle über das Geschehen im Klassenzimmer an die Schülerinnen und Schüler ab (z. B. lässt Lernende etwas erklären, stellt Fragen, geht auf Beiträge ein)	Die Schülerinnen und Schüler reagieren auf Aufforderungen der Lehrperson, zeigen jedoch wenig Eigeninitiative (z. B. erledigen Aufträge, beantworten Fragen, hören zu)
2	Die Lehrperson zeigt kaum Verhalten, mit dem das Geschehen im Klassenzimmer beeinflusst oder kontrolliert werden könnte. Sie ist zurückhaltend (z. B. inkonsequent, führt kaum)	Die Schülerinnen und Schüler reagieren lediglich auf direkt an sie gerichtete Aufforderungen und zeigen keine Eigeninitiative (z. B. reagieren kaum, nur wenn sie explizit aufgerufen werden, sind abwartend, antworten leise oder ängstlich)
1	Die Lehrperson verhält sich sehr passiv. Sie zeigt kein Verhalten, mit dem das Geschehen im Klassenzimmer beeinflusst oder kontrolliert werden könnte (z. B. interveniert nicht)	Die Schülerinnen und Schüler verhalten sich sehr passiv (z. B. beantworten Fragen nicht, wirken verängstigt)

Table 10.2 Ranking system for the two indicators of the sub-dimension *communion* (Scherzinger et al. 2020)

	Lehrperson	Schülerinnen und Schüler
5	Die Lehrperson zeigt Wärme und ist freundlich, jedoch distanzlos (z. B. wahrt professionelle Distanz nicht, verhält sich kumpelhaft, macht Sprüche oder Witze)	Die Schülerinnen und Schüler sind herzlich oder freundlich, jedoch kumpelhaft und distanzlos gegenüber der Lehrperson (z. B. umarmen oder berühren sie, verhalten sich kumpelhaft)
4	Die Lehrperson zeigt Wärme und ist freundlich und wahrt die professionelle Distanz (z. B. lobt oder unterstützt Lernende, zeigt Wertschätzung, nimmt sich Zeit)	Die Schülerinnen und Schüler sind herzlich oder freundlich, wahren aber eine angemessene Distanz gegenüber der Lehrperson (z. B. unterstützen die Lehrperson, freuen sich, bedanken sich)
3	Die Lehrperson verhält sich neutral, sie ist weder freundlich noch gereizt oder unfreundlich (z. B. spricht in einem neutralen Ton, reagiert oder antwortet recht sachlich und kurz)	Die Schülerinnen und Schüler sind neutral, weder freundlich noch unfreundlich oder gereizt gegenüber der Lehrperson (z. B. zeigen wenig positive oder negative Emotionen, verhalten sich respektvoll)
2	Die Lehrperson ist kühl, unfreundlich oder distanziert (z. B. reagiert gereizt, unfreundlich, droht, unterbricht Lernende)	Die Schülerinnen und Schüler wirken abweisend, reserviert, unfreundlich, provokativ oder distanziert gegenüber der Lehrperson (z. B. ignorieren die Lehrpersonen, machen sich lustig, sind verängstigt, gehen wenig auf die Lehrperson ein)
1	Die Lehrperson ist feindselig gegenüber den Lernenden (z. B. beschimpft oder beleidigt oder wird handgreiflich)	Die Schülerinnen und Schüler verhalten sich feindselig gegenüber der Lehrperson (z. B. verhalten sich aggressiv, widersetzen sich aktiv, werden handgreiflich)

10.3   Results and Comparison

In this study, only the results of the global values for the indicators are examined. The categorisation into relation types is not part of this study as in Section 9.3. This is because only the reasons for the differences in the results of the two methods will be examined.

The estimates for the overall results of the indicators are compared using figures such as Fig. 10.2.

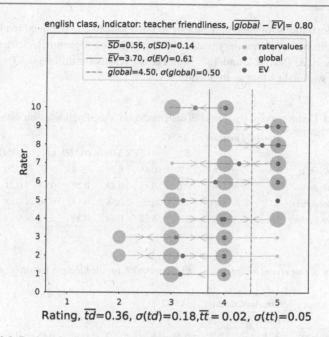

Figure 10.2 Ratings for indicator teacher friendliness in English class with 23 raters

This Figure 10.2 is similar to the interpretation of global estimations in Section 7.3. In addition to this type of figure, the global ratings made by the students before the ASSG method indicators were evaluated are also included. As with the expected value, the global ratings are also calculated in a mean value.

The absolute difference between the mean of the global ratings $\overline{global}$ and the mean value of the expected values $\overline{EV}$ is also calculated. This difference can be found in the title of the plots. A high difference, such as $|\overline{global} - \overline{EV}| = 0.80$, indicates that the standard method and the ASSG method result in different values. How big the difference is for the respective indicator can be estimated using this term.

Additionally to the mean values of the other parameters of the ASSG method (EV, SD, and td), their standard deviation is also calculated. For example, $\sigma(EV)$ indicates how much the expected values of each rater vary. A low value for $\sigma(EV)$ means that the expected values for each rater vary only slightly. In contrast, a high value for $\sigma(SD)$ indicates that the variability of each rater's scores varies greatly. This means that one rater, for example, might only give the same rating, while another rater varies between all possible ratings.

The illustrations for each indicator in all rated lessons can be found in chapter G of the electronic supplementary material. The results for all parameters are summed up in Tab. 10.3, Tab. 10.4, and Tab. 10.5. Higher results for differences (> 0.3) are indicated with italics font as in Section 9.3.

Table 10.3 Comparison of the results in both methods for the English lesson with 23 raters

indicator	standard method		ASSG						Δ
	global	σ	EV	σ(EV)	SD	σ(SD)	td	σ(td)	
teacher activeness	4.3	0.64	3.3	0.45	0.73	0.24	0.57	0.29	*1.00*
student activeness	3.5	0.92	3.1	0.45	0.83	0.29	0.49	0.21	*0.40*
teacher friendliness	4.5	0.50	3.7	0.61	0.56	0.14	0.36	0.18	*0.80*
student friendliness	3.9	0.83	3.4	0.52	0.49	0.30	0.31	0.24	*0.50*

Table 10.4 Comparison of the results in both methods for the German lesson (poems) with 9 raters

indicator	standard method		ASSG						Δ
	global	σ	EV	σ(EV)	SD	σ(SD)	td	σ(td)	
teacher activeness	3.78	0.92	3.44	0.34	0.75	0.19	0.48	0.17	*0.33*
student activeness	3.33	0.67	3.44	0.32	0.91	0.23	0.59	0.18	0.11
teacher friendliness	3.78	1.13	3.67	0.68	0.46	0.22	0.25	0.18	0.11
student friendliness	4.00	0.82	3.44	0.66	0.56	0.25	0.31	0.17	*0.56*

Table 10.5 Comparison of the results in both methods for the German lesson (Schiller) with 16 raters

indicator	standard method		ASSG						Δ
	global	σ	EV	σ(EV)	SD	σ(SD)	td	σ(td)	
teacher activeness	3.93	0.77	3.33	0.45	0.71	0.23	0.49	0.21	*0.60*
student activeness	3.60	0.61	3.47	0.39	0.76	0.32	0.48	0.24	0.13
teacher friendliness	4.47	0.50	4.07	0.31	0.53	0.15	0.33	0.12	*0.40*
student friendliness	4.00	0.63	3.60	0.35	0.42	0.23	0.30	0.20	*0.40*

As can be seen from the tables above, there are large differences between the results of the standard method and the ASSG method for most indicators. The number of people rating the indicators does not seem to lead to a smaller difference. For example, the ratings of the indicators for the English lesson all show a big difference in their results. This is despite the fact that this lesson was rated by 23 people. In contrast, the lesson with the fewest raters also has the smallest difference in results.' Thus, the possibility that the difference in methods is due to the small number of raters in the last study does not seem to be the cause of this difference.

A high SD for the rating or a higher travel distance do not seem to be the cause either. For example, the indicator *student activeness* has a small difference in the results, but also a high SD and td (see Fig. 10.3). The differences in the parameters of the ASSG method from rater to rater also do not seem to influence this difference. This is the case because the standard deviations for the parameters (SD and td) do not differ much from each other.

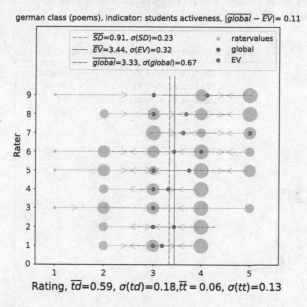

Figure 10.3 Ratings for indicator student activeness in German class (poems) with 9 raters

However, one thing is noticeable in the tables. The standard deviation for the global ratings is always higher than the standard deviation for the expected values, except in one case. This indicates that the global ratings of the raters differ more from rater to rater than the expected values of their ratings for the ASSG method. This means that a new rater that has not yet rated the indicator is more likely to reflect the same expected value as the other raters before. It is also less likely that a new rater will give the same global rating. Therefore, the subjective influence on indicator ratings seems to be greater for global ratings than for expected values. Thus, the expected values are more objective than the global ratings, which leads to different results. Consequently, the inter-coder reliability is higher for the ASSG method than for the standard method.

One reason for this effect is that global ratings are more extreme at positive or negative ratings than ratings at short intervals. This can be seen in Fig. 10.4. In this case, the global ratings are almost always the best ratings of the variable ratings of the ASSG method. For example, the sixth rater has given ratings between 2 and 5, but his global rating is 5. This effect is even greater for the eighth rater, who has given ratings in time intervals 3 and 4, but his global rating is also 5. This means that global ratings are rated higher than ratings of individual time intervals. The results of the two methods are therefore different.

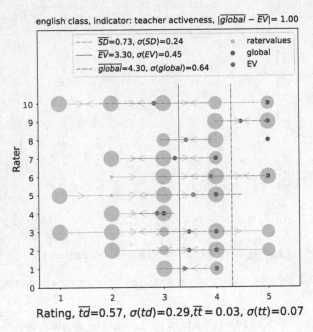

Figure 10.4 Ratings for indicator teacher activeness in English class with 23 raters

Another reason for the different results is that the individual events have a higher influence in the standard method than in the ASSG method. This can be explained by the fact that the ratings in the standard method are carried out by selecting specific events on which to base the rating. In the ASSG, such specific events only lead to a single rating for one of the many time intervals. Therefore, the influence of such single events is much smaller in the ASSG method than in the standard method.

The dynamics of the ratings also have an influence in the ASSG method, as the expected value is influenced by all ratings, even if such a rating is only carried out once. With global ratings, such dynamics cannot be taken into account and quantified. In the ASSG method, the researcher also has two other parameters at his disposal that can describe how much the ratings differ over the duration of the lesson. Such parameters do not exist in the standard method. The dynamics contained in the expected value therefore also lead to differences between the two methods.

In summary, the differences in the results of the two methods are not caused by a low number of raters, as there are still high differences in this study with 23 raters. The difference is also not caused by the fact that the global ratings and the ASSG ratings are carried out by different people, as this is not the case in this study. Rather, these differences are due to fundamental differences in the two methods. It is not possible to say which method leads to values that better represent the indicator rating of teaching, as there is no "correct" result. However, it can be seen that the objectivity of the ASSG method is higher than that of the standard method. In addition, the ASSG method has more opportunities for knowledge, which also describe how representative the global rating of the indicator value is.

Application of Advanced State Space Grids to the analysis of physics teaching

11

In this chapter, the final research question "How can the ASSG method be used for the analysis of physics teaching?" is addressed. The application of the ASSG method will be divided into general applications and specific applications. The two specific applications that are of particular interest for the analysis of physics teaching are dealt with in Section 11.3 and Section 11.4.

11.1 General Applications

In Chapter 9 and Chapter 10, the ASSG method was applied on German and English lessons. The general application of the ASSG method does not differ between the subjects studied. For example, the aspects of the basis dimensions are applicable to all teaching subjects (Lipowsky 2019). Therefore, the ASSG method can also be used to examine general aspects of teaching in physics classes. The requirements for the indicators that apply to the ASSG method also do not change in physics lessons. The general application of the ASSG method therefore does not change when analysing physics teaching.

The possible topics in the analysis of physics teaching that can be investigated using the ASSG method can be different. Since this method can investigate the dynamic progression of aspects of physics teaching, it can be used for aspects that are unique in the analysis of physics teaching.

N. Litzenberger, *Introduction of Advanced State Space Grids and Their Application to the Analysis of Physics Teaching*, BestMasters, https://doi.org/10.1007/978-3-658-42732-0_11

11.2   Dynamic Interaction with Worksheets and Student Experiments

Student experiments are an essential part of physics lessons. In these experiments, students work independently on a scientific question by formulating hypotheses, testing these hypotheses through experiments, and interpreting the results in relation to their hypotheses (Klahr et al. 1989). In doing so, students have to overcome numerous hurdles in their learning process, such as the metacognitive requirement to conduct an experiment themselves, which can overwhelm some students (de Jong et al. 1998). To counteract these problems, worksheets are used to help students step by step in their learning process (Lunetta 1998). These problems of the students cannot be generalised in the basic dimensions and require subject-specific adaptations for physics teaching (Praetorius et al. 2020).

In order to investigate this learning process of the students, field experiments are often used for studies (eg. Wirth et al. 2008). However, these have very limited access to new insights (ibid.), as they cannot map the dynamic learning process of the students. The study of the dynamic learning process can be further investigated using the ASSG method.

In particular, the successive impact of the worksheets and their tasks on the learning process can be depicted and examined with the ASSG method. If a certain task is not suitable for the subjects, this would lead to a deterioration in the ranking of certain indicators. Such indicators could, for example, describe the students' workflow and whether they get stuck on a task. How helpful the worksheet was designed for the students can then be investigated with the ASSG method.

For example, an indicator such as "The students follow their work assignment." could measure the workflow by counting the students who follow the work assignment. In this particular case, one needs to define what counts as an action that follows the work assignment. It is for example unclear, whether a student that is looking at a wall can be counted as student that is working or not. However, it remains unknown if the person in question is actually thinking about the assignment or not. Another indicator to measure the workflow can be determined by tracking eye movement (eg. Pellicer-Sánchez et al.). This can be used to measure how long students read a text on the worksheet and whether they jump between lines or figures. In this way, another aspect of the workflow can be measured.

If the students get stuck in a certain part of the work assignment, they cannot actively follow it. This would lead to a lower rating of the indicators. Therefore, the points in the ASSG plot will leave a stable attractor and change their rating. If students are only stuck for a short period of time, the points in the ASSG plot will return to the attractor. When short periods of student being stuck are frequent, the ASSG plot displays a high middle standard deviation and travel distance. This is because the workflow indicators change more frequently.

If the students were to get stuck for a longer period of time because they encountered a major problem, their workflow would change. This leads to a new, different attractor. This can also be measured with the ASSG method by using two different expected values and deviation curves. The difference between the two expectation values indicates how big the difference between the workflows is. The middle standard deviations and travel distances can be used to measure whether students have short changes in workflow when solving their problem.

Thus, the ASSG method can measure the stability and the disturbances of the workflow with numerical parameters and thus the interaction with the experiment. Through these parameters, the lessons and the worksheets can be compared and it can be evaluated which kind of worksheet tasks have the best influence on the interaction between worksheet and experiment.

11.3 Influences of Single Events on the Workflow

Another possible application of the ASSG method in the analysis of physics teaching is to measure the influence of single events that occur only once in the lesson on the students' workflow. An example of such a single event could be how the teacher starts a new lesson sequence and motivates its content. This is an important aspect of teaching physics as students' interest in science is declining (Djudin 2018). Another example could be how the teacher explains the task at hand and whether they forget to explain essential aspects of the task. Such incomplete explanations can lead to unresolved questions in the lesson (Viennot 2021).

Similar to the interaction between worksheet and experiment, indicators can measure the students' workflow with the ASSG method. If the teacher starts a new teaching sequence in a way that is less understandable for the students, an impact on the students' workflow can be detected by rating the indicators. If the workflow is disturbed because a key moment of the task has not been explained, the rating of the indicators is likely to change. This is similar to the influence of worksheets on the workflow and interaction with the experiment. In the same way as explained in Section 11.2, the stability and change of the workflow can be measured with the ASSG method.

In addition to measuring the workflow, the explanation itself can also be measured. For example, the non-verbal aspects of an explanation can be measured if these aspects can be assessed sufficiently often. As explained in Section 8.2, ten ratings are barely enough to apply the ASSG method. This means that an explanation that lasts five minutes and is rated every 30 seconds by the indicators can be measured with the ASSG method. Shorter explanations would have to be evaluated differently, e.g. by rating an indicator only once.

Such non-verbal aspects of an explanation can include body movements such as gestures or posture, para-linguistic features such as voice quality or non-linguistic sounds such as laughing or yawning, and the use of artefacts (Duncan 1969). All these aspects can be measured with the ASSG method. For example, voice qualities can be measured by assessing how loud each part of the sentence was. This can be assessed by measuring the loudness of the voice. The ASSG method can then measure how much the loudness of the voice changes with the travel tendency and the standard deviation. The average loudness of the voice is indicated by the expected value. In addition, the relationships between the loudness and other non-verbal aspects can be assessed with the relation types.

How the aspects of the explanation influence the workflow can then be examined by combining the ASSG for the aspects of the explanation with the ASSG plot of the workflow. In this way, the influence of individual events, such as the explanation of the task, on the workflow can be examined through two different ASSG plots. In addition, it is possible to evaluate the acquired knowledge with evaluation sheets or a test after the lesson.

With this idea of evaluating individual events differently and combining them with an ASSG plot of dynamic aspects, studies can also be conducted outside of physics lessons. For example, the influence of non-verbal communication on the success of a sales talk can be studied in the same way. The ASSG method can measure the different aspects and dynamic changes of non-verbal communication. For example, non-verbal communication can be evaluated where the voice quality changes frequently and the use of gestures remains constant. These differences in non-verbal communication can then be linked to the success of the sales conversation. Thus, questions such as "Does the dynamic change between different types of nonverbal communication influence the success of a sales talk?" can be investigated using the ASSG method.

11.4   Potential in Disruption Intervention

Dealing appropriately with disruptions in physics lessons is an important part of good physics teaching (Merzyn 2015). Especially in experimental lessons, where many students do experiments themselves, good classroom management is important (Praetorius et al. 2020). Good classroom management can be disrupted by the actions of a single student. To prevent such disruptions, active and proactive interventions by the teacher can contribute to a more stable classroom management (Zuckerman 2007). Proactive interventions can prevent further disruptions. For example, a change of pace (ibid.) or student participation, among other things, can also prevent classroom management disruptions (Klaffke 2020).

Another type of interventions are active interventions (Zuckerman 2007). These are interventions that are directly directed at an occurring disruption. These direct interventions range from non-verbal glances at the students all the way to an exclusion from the lesson (Klaffke 2020). The aim of these interventions is to quickly address the disruption and restore stable classroom management.

However, these intervention possibilities of the teacher are often argued with the experiences of the authors. For example, Klaffke (2020) argues the effects of individual interventions with his own experiences or generalised arguments. However, arguments based on empirical findings and measurements of the effect size of such interventions are not often used.

A study to measure the effect size of each active intervention can be conducted using the ASSG method. As explained in Section 7.4, the ASSG method can measure the difference between an attractor before a disturbance occurs and the difference between the attractor after the disturbance. In doing so, the effect of an intervention can be measured by comparing the expected value and the deviation curve of the stability before and after the disturbance.

Indicators that describe student workflow and classroom management can be used to describe the stability of classroom management with the ASSG method. A low variability of these indicators suggests that classroom management is stable. This means that the ratings of the indicators that measure the aspects of classroom management remain the same. These aspects of classroom management therefore remain stable and do not change. In contrast, high variability indicates low stability of classroom management. This is due to the fact that when the indicators are rated differently, the aspects of classroom management and work processes also vary. Thus, a disruption in classroom management also affects the ratings of these indicators. After the disturbance caused by the intervention has been resolved, the indicator values return to a stable situation. This new stability can then be interpreted as a new attractor with its own expected value and deviation curve.

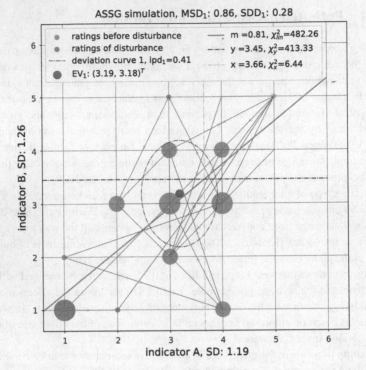

Figure 11.1 ASSG for the effect size of disruption interventions without ratings after the disruption

An example of an ASSG plot up to the point of the disruption can be seen in Fig. 11.1. In this case, the ratings for both indicators are almost exclusively between two and four until the disruption occurs. These ratings are indicated by a dark blue colour in the ASSG plot. When the disruption occurs, the ratings of the indicators decrease. These ratings are shown as red circles in the ASSG plot.

A disruption such as the one in Fig. 11.1 takes a long time to be resolved. How many ratings can be made depends on the time resolution of the ratings and the duration of the disruption. Imagine a student refuses to do his assignment and starts arguing with the teacher. The resolution of such a disruption could take several minutes. During the duration of the disruption, a rater can rate each indicator every 60 seconds. This results in four ratings if the disruption takes four minutes to be resolved. Such a disruption can be represented by an ASSG plot.

Shorter disruptions cannot be shown in an ASSG plot if the number of ratings per minute is not adjusted. Such short disruptions can occur, for example, when two students argue with each other and the teacher looks at them to get them to stop. This interaction may last as little as 20 seconds. In a rating where each indicator is rated every 60 seconds, such an interaction cannot be shown in detail in an ASSG plot. Instead, the rating is changed slightly during these 60 seconds, resulting in a higher MSD in the ASSG plot. If a researcher wants to examine such short disturbances in an ASSG plot, the indicators would have to be rated more frequently. In this case, a rating would be required every ten seconds or even more often. This is generally not a problem with the ASSG method, as more ratings lead to a better representation of the EV. However, it is not easy for a human rater to perform a rating every ten seconds. In these cases, it may be more appropriate to use the ASSG method for the short duration of the disturbance. This is similar to the analysis of a teacher's non-verbal actions in Section 11.3.

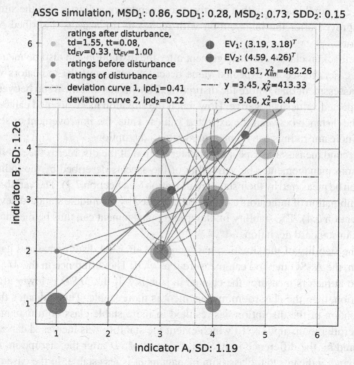

Figure 11.2 ASSG for the effect size of disruption interventions

In the case of a longer duration of a disruption, the rating returns to a stable attractor after the disturbance has been resolved. This situation can be rated with the indicators again and applied to the ASSG plot. These new ratings are indicated by a light blue colour. This is due to the fact that the difference between the ratings before and after the disruption can be shown in the ASSG plot. Such an ASSG plot can be seen in Fig. 11.2. This is the ASSG plot which was already mentioned in Section 7.4. In the case of fig 11.2, the ratings after the disruption form a new attractor. The differences between the pre- and post-disruption ratings can be seen in the ASSG plot. This shift in stabilities can be measured by the parameters of the ASSG method as follows.

If the expected value of the second attractor improves as in the example of Fig. 11.2, the travel tendency between the two expected values is $tt_{EV} = +1.00$. The evaluation of the indicators thus improves after the intervention of the disruption. How great this improvement is can be measured by the distance between the two expected values, which in this case is $td_{EV} = 2.16$. A higher value of this travel distance in combination with a positive travel tendency indicates that the situation has improved after the intervention with regard to the aspects described by the indicators.

A setback in classroom management after the disruption can also be measured. In this case, the second expected value describes ratings of the indicators with a worse assessment than before the disruption. This means that the trend between the two expected values is negative. How much the situation has worsened can be seen from the distance between the expected values. Thus, the improvement or setback in the indicator ratings can be measured after a disruption.

A second measurable aspect of the intervention of the disorder is the stability of classroom management. The variability of the aspects described by the indicators used can be measured by their standard deviation (see Section 7.3). The variability of the combination of indicators can then be indicated by the middle standard deviation (see Section 6.4). The stability of classroom management can thus be measured by the mean standard deviation.

If the stability of the situation has increased after the disturbance and its intervention, the ASSG method can measure it as well. The difference in the MSDs of the two attractors measures the change in stability. If the MSD is lower after the disruption, then the classroom management is more stable. This indicates that the intervention of the disruption has resulted in more stable classroom management than before the disruption. To which extent the stability was increased can also be measured by the difference in MSDs. A lower MSD after the disruption, on the other hand, indicates that classroom management is less stable. In the case of Fig. 11.2, the difference between the two MSDs is 0.13, with the second MSD being lower. Thus, the situation in the classroom is more stable after the intervention of the disruption.

In summary, the ASSG method can measure two different effect sizes of the proactive intervention of a disruption. The first is the increase or decrease in the average ratings of the indicators. This indicates whether the measured aspects of classroom management or workflow have improved or worsened. The second effect size is the increase or decrease in the stability of classroom management after the intervention. If the ratings of the indicators are frequent enough it is also possible to measure the interaction of the disruption itself.

In this way, the effect sizes of possible proactive interventions can be measured. A study using the ASSG method can then rank different proactive interventions in terms of their effect sizes. This would then make it possible to create a ranking list of the best disruptive interventions and substantiate it with empirical evidence.

Conclusion and Outlook

12

In this thesis it was shown how the State Space Grid method can be enhanced to the Advanced State Space Grid method by defining new parameters such as the expected value of an ASSG plot. This lead to a categorisation of different relation types. The critical parameter values to distinguish between the relation types were simulated. The results of this simulation lead to a flow chart with which a first categorisation of ASSG plots can be successfully performed.

This new ASSG method was tested in two different studies. In the first study, it was shown that a successful categorisation of the different types of relations is possible in actual lessons. Thus, an application of the ASSG method for the analysis of dynamic interactions in physics lessons is possible. However, the results of the ASSG method differ in comparison to an standard method. This difference can be caused by the fundamental differences between these methods.

An application of the ASSG method for the analysis of specific aspects of physics teaching is also possible. An example of this is the study of students' dynamic interaction with worksheets and experiments. Research gaps such as the relationship between basic and sub-dimensions of good teaching as well as effect sizes of proactive disruption interventions can also be measured with the ASSG method.

However, future studies need to test the use of the ASSG method in measuring the effect size of interruption interventions. In addition, a comparison between the thin-slicing method and the ASSG method may be interesting as they share the similarity of data collection of indicators that are rated at fixed time intervals.

N. Litzenberger, *Introduction of Advanced State Space Grids and Their Application to the Analysis of Physics Teaching*, BestMasters, https://doi.org/10.1007/978-3-658-42732-0_12

A different application of the ASSG method, not covered in this paper, are correlation studies. It is possible to apply global estimates of the indicator ratings of a single lesson to the ASSG plot. If different lessons of the same teacher and the same class are rated, the dynamic progression of the global ratings of the indicators during different lessons can be studied. In this way, it can be investigated whether the specific aspects of the indicators improve over the course of the school year or whether the aspects of the indicators remain stable during this period.

The ASSG method has room for further improvement though. The critical values of the flow chart can be improved by simulating the critical value close to zero at which the flow chart results in a stationary or linear relation. As it stands, the flow chart only tests if $\chi^2 \approx 0$. How low the χ^2 value has to be for an ASSG plot in order to definitely result in a stationary or linear relation and no other type of relation remains unknown in this thesis.

The flow chart can also be improved by including different sizes in the indicator rankings. In the current state of the flow chart, only indicator combinations with five different possible ratings are possible. In order to adapt the flow chart for indicators with only four possible ratings, the simulations have to be adapted as well. The same applies to indicators with more than 5 possible ratings. For each different size in the ranking, each simulation must be adjusted.

In Section 6.1 the lack of uniqueness of the points in the ASSG plot was discussed. In future improvements of the ASSG method this aspect could be addressed. A possible way to achieve the uniqueness of the points would be to represent the time increase by a dark colour of the circle around the point in the grid. A similar approach was taken in the example of the ASSG plot to measure the effect size of an intervention in the case of a disruption in fig. 11.2. In this case, the post-interruption rating was plotted with a brighter blue than the pre-interruption rating. The dynamic increase in the time during which the indicator rating was carried out can be shown in a similar way. One possibility would be to start the first rating with a very bright blue colour and then successively decrease the brightness of the circle around the grid location.

The problem of time resolution has to be addressed as well. As mentioned in Section 11.4, short disturbances that only last a few seconds can barely be shown in an ASSG plot. An analysis of such interactions with the ASSG method relies on indicators that have to be rated very frequently. Such frequent ratings can hardly be made by a human being. Instead, a rating by physically measurable aspects can solve this problem. An example of such an aspect is the measurement of noise. However, such indicator ratings are limited to certain aspects of lesson analysis. This leads to a further reduction of usable indicators for the ASSG method.

An other improvement for the ASSG method is the addition of more than two combinations of indicators. For example, it is possible to display all parameters in a three-dimensional ASSG plot. This means that the ASSG method can be expanded to analyse dynamic interactions between n indicators. The different relation types can also be adjusted to an n dimensional ASSG plot.

Since the ASSG method is still new and has only been tested in two small studies, further research is needed to gain better insight into this new method. The objective of future studies will be to investigate further applications of this new method.

Bibliography

Ambady, N., Bernieri, F. J., & Richeson, J. A. (2000). Toward a histology of social behavior: Judgmental accuracy from thin slices of the behavioral stream. In Advances in experimental social psychology (Vol. 32, pp. 201–271). Academic Press.

Ambady, N., & Gray, H. M. (2002). On being sad and mistaken: Mood efects on the accuracy of thinslice judgments. Journal of Personality and Social Psychology, 83, 947–961.

Ambady, N., & Rosenthal, R. (1993). Half a minute: Predicting teacher evaluations from thin slices of nonverbal behavior and physical attractiveness. Journal of Personality and Social Psychology, 64, 431–441.

Adelman, H.S. & Taylor, L. (2005). Classroom climate. In S.W. Lee (Hrsg.), Encyclopedia of school psychology (S. 88–90). Thousand Oaks: Sage.

Bär, C. (2010). Elementare Differentialgeometrie. de Gruyter.

Baumert, J., Kunter, M., Blum, W., Brunner, M., Voss, T., Jordan, A., Klusmann, U., Krauss, S., Neubrand, M. & Tsai, Y.-M. (2010). Teachers' mathematical knowledge, cognitive activation in the classroom, and student progress. American Educational Research Journal, 47, 133–180.

Beetz, J. (2015). Kartesische Koordinaten. In Funktionen für Höhlenmenschen und andere Anfänger (pp. 3–19). Springer Spektrum.

Behrends, E. (2012). Elementare Stochastik: Ein Lernbuch-von Studierenden mitentwickelt. Springer-Verlag.

Berendsen, H. J. (2011). A student's guide to data and error analysis. Cambridge University Press.

Birkner, M. (2021): Einführung in die Stochastik. Notizen zu einer Vorlesung an der Johannes-Gutenberg-Universitat Mainz, Winter 2020/2021. https://www.staff.uni-mainz.de/birkner/GrundlStoch_2021/Stochastik-Einfuehrung_WS2021.pdf.

Blaikie N. (2003): Analyzing Quantitative Data London: SAGE Publications Ltd.

Bohl, T. & Kucharz, D. (2010). Offener Unterricht heute. Konzeptionelle und didaktische Weiterentwicklung. Beltz Verlag.

Bortz, J., & Döring, N. (2016). Forschungsmethoden und Evaluation für Human- und Sozialwissenschaftler. 3. Überarbeitete Auflage. Springer.

Bovier, A. (2019): Einführung in die Wahrscheinlichkeitstheorie. Vorlesung Winter 2019/20, Bonn. https://www.dropbox.com/s/wllhl2v4ccv2x7f/wt-new.pdf?dl=0.

Brophy, J. (2000). Teaching. Brussels: International Academy of Education.

Bylieva, D., Lobatyuk, V., Safonova, A., & Rubtsova, A. (2019). Correlation between the Practical Aspect of the Course and the E-Learning Progress. Education Sciences, 9(3), 167.

Çalik, S., & Güngör, M. (2004). On the expected values of the sample maximum of order statistics from a discrete uniform distribution. Applied mathematics and computation, 157(3), 695–700.

Cochran, W. G. (1952). The χ^2 test of goodness of fit. The Annals of mathematical statistics, 315–345.

Cohen L, Manion L & Morrison K. (2000): Research Methods in Education. 5th edn. RoutledgeFalmer.

Chatti, M. A., Lukarov, V., Thüs, H., Muslim, A., Yousef, A. M. F., Wahid, U., ... & Schroeder, U. (2014). Learning analytics: Challenges and future research directions. eleed, 10(1).

Clark, A., Gilmore, S., Hillston, J., & Tribastone, M. (2007). Stochastic process algebras. In International School on Formal Methods for the Design of Computer, Communication and Software Systems, 132–179. Springer Spektrum.

Ditton, H. (2008): Qualitätsvolles Lehren und Lernen. In: Wiater, Werner & Pötke, Regina (Hrsg.): Gymnasien auf dem Weg zur Exzellenz. Wie lässt sich Qualität am Gymnasium entwickeln?, 54–62.

Djudin, T. (2018). How to cultivate students' interests in physics: A challenge for senior high school teachers. Jurnal Pendidikan Sains, 6(1), 16–22.

Duncan Jr, S. (1969): Nonverbal communication. Psychological Bulletin, 72(2), 118.

Döring, N. & Bortz, J. (2016): Forschungsmethoden und Evaluation in den Sozial- und Humanwissenschaften.

Döppl, Sabrina (2003). Körpersprache des Lehrers im Unterricht. In: GRIN Wissen finden und publizieren, https://www.grin.com/document/108086.

Dyckhoff, A. L., Zielke, D., Bültmann, M., Chatti, M. A., & Schroeder, U. (2012). Design and implementation of a learning analytics toolkit for teachers. J. Educ. Technol. Soc., 15(3), 58–76.

Eder, F. (2002). Unterrichtsklima und Unterrichtsqualität. Unterrichtswissenschaft, 30(3), 213–229.

Eichler, A., & Vogel, M. (2011). Leitfaden Stochastik. Vieweg+ Teubner.

Evertson, C. M. & Weinstein, C. S. (Eds.) (2006). Handbook of Classroom Management. Research, Practice, and Contemporary Issues. Mahwah, N.J: Lawrence Erlbaum Associates.

Fornasini, P. (2008). The uncertainty in physical measurements: an introduction to data analysis in the physics laboratory (Vol. 995). Springer.

Fowler, K. A., Lilienfeld, S. O., & Patrick, C. J. (2009). Detecting psychopathy from thin slices of behavior. Psychological Assessment, 21, 68–78.

Gabriel, K. (2014). Videobasierte Erfassung von Unterrichtsqualität im Anfangsunterricht der Grundschule. Klassenführung und Unterrichtsklima in Deutsch und Mathematik. Kassel: Kassel University Press.

Glahn, C. (2009). Contextual support of social engagement and reflection on the Web. Open Universiteit, https://www.ou.nl/documents/40554/111697/Christian_Glahn_thesis_2009.pdf/a3baceb1-e3e1-4f1f-a1f1-75ce55359158

Grinstead, C., & Snell, L. J. (2006). Introduction to probability.

Gruehn, S. (2000). Unterricht und schulisches Lernen. Schüler als Quellen der Unterrichtsbeschreibung. Waxmann.

Goh, J. X., Ruben, M. A., & Hall, J. A. (2019). When social perception goes wrong: Judging targets' behavior toward gay versus straight people. Basic and Applied Social Psychology, 41, 63–71.

Hable, R. (2015). Einführung in die Stochastik: ein Begleitbuch zur Vorlesung. Springer.

Hattie, J. (2009): Visible Learning. A synthesis of over 800 meta-analyses relating to achievement.

Helmke, A. (2006). Was wissen wir über guten Unterricht? Über die Notwendigkeit einer Rückbesinnung auf den Unterricht als dem „Kerngeschäft"von Schule. Pädagogik, 2, 42–45.

Helmke, A. (2007). Aktive Lernzeit optimieren. Was wissen wir über effiziente Klassenführung? Pädagogik, 5, 44–48.

Helmke, A. (2012). Unterrichtsqualität und Lehrerprofessionalität. Diagnose, Evaluation und Verbesserung des Unterrichts (6. Aufl.). Seelze-Velber: Kallmeyer.

Henze, N. (1997). Stochastik für Einsteiger (Vol. 8). Vieweg.

Hirschmann, N., Kastner-Koller, U., Deimann, P., Schmelzer, M., & Pietschnig, J. (2018). Reliable and valid coding of thin slices of video footage: Applicability to the assessment of mother-child interactions. Journal of Psychopathology and Behavioral Assessment, 40, 249–258.

Hollenstein T. (2007): State space grids: Analyzing dynamics across development. International Journal of Behavioral Development;31(4): 384–396.

Hollenstein, T. (2013). State space grids. In State space grids. Springer.

Houser, M. L., Horan, S. M., & Furler, L. A. (2007). Predicting relational outcomes: An investigation of thin slice judgments in speed dating. Human Communication, 10, 69–81.

Huth, K. (2004): Entwicklung und Evaluation von fehlerspezifischem informativem tutoriellem Feedback (ITF) für die schriftliche Subtraktion. http://www.qucosa.de/fileadmin/data/qucosa/documents/1243/1105354057406-4715.pdf

Imhof, M & Bellhäuser, H (2021): Psychologische Forschungsmethoden in den Bildungswissenschaften. Eine Einführung für Lehramtsstudierende. Hogrefe Verlag.

Jamieson, s. (2004): Likert scales: how to (ab)use them? Medical Education, 38(12), 1217–1218.

Jentsch A., Casale, G., Schlesinger, L., Kaiser, G., König, J., Blömeke, S. (2019). Variabilität und Generalisierbarkeit von Ratings zur Qualität von Mathematikunterricht zwischen und innerhalb von Unterrichtsstunden. Springer Fachmedien.

de Jong, T., & van Joolingen, W. R. (1998). Scientific discovery learning with computer simulations of conceptual domains. Review of educational research, 68(2), 179–201.

Klaffke, T. (2020): Unterrichtsstörungen—Präventionen und Intervention. Möglichkeiten einer ressourcenorientierten Pädagogik. Klett.

Klahr, D., Dunbar, K., & Fay, A. L. (1989). Designing good experiments to test bad hypotheses. CARNEGIE-MELLON UNIV PITTSBURGH PA ARTIFICIAL INTELLIGENCE AND PSYCHOLOGY PROJECT.

Klippert, H. (2008). Besser lernen. Kompetenzvermittlung und Schüleraktivierung im Schulalltag. Klett.

Klenke, A. (2020): Wahrscheinlichkeitstheorie (4. edition). Springer Spektrum.

Klieme, E. & Rakoczy, K. (2008). Empirische Unterrichtsforschung und Fachdidaktik. Outcome-orientierte Messung und Prozessqualität des Unterrichts. Zeitschrift für Pädagogik, 54(2), 222–237.

Klieme, E./Schümer, G./Knoll, S. (2001): Mathematikunterricht in der Sekundarstufe I: „Aufgabenkultur" und Unterrichtsgestaltung im internationalen Vergleich. In: Klieme, E./Baumert,J. (Hrsg.): TIMSS—Impulse für Schule und Unterricht. BMBF, 43–57.

Kobarg, M. & Seidel, T. (2003). Prozessorientierte Lernbegleitung im Physikunterricht. In T. Seidel, M. Prenzel, R. Duit & M. Lehrke (Hrsg.), Technischer Bericht zur Videostudie „Lehr-Lern-Prozesse im Physikunterricht", 151–200. IPN.

Kounin, Jakob S. (2006): Techniken der Klassenführung. Band 3, Waxmann.

König, J. & Pflanzl, B. (2016). Is teacher knowledge associated with performance? On the relationship between teachers' general pedagogical knowledge and instructional quality. European Journal of Teacher Education, 39, 419–436.

Kraus, M. W., & Keltner, D. (2009). Signs of socioeconomic status: A thin-slicing approach. Psychological Science, 20, 99–106.

Kuzon, Jr. WM/Urbanchek MG (1996): & McCabe S. "The seven deadly sins of statistical analysis" Annals Plastic Surg, 37: 265–272.

Levine, S. P., & Feldman, R. S. (1997). Self-presentational goals, self-monitoring, and non-verbal behavior. Basic and Applied Social Psychology, 19, 505–518.

Li, S., Ogura, Y., & Kreinovich, V. (2013). Limit theorems and applications of set-valued and fuzzy set-valued random variables (Vol. 43). Springer Science & Business Media.

Lipowsky, F. & Bleck, V. (2019). Was wissen wir über guten Unterricht. Ein Update. In: Steffens, U. & Messner, R. (Hrsg.): Unterrichtsqualität. Konzepte und Bilanzen gelingenden Lehrens und Lernens.

Lipowsky, F. & Hess, M. (2019). Warum es manchmal hilfreich sein kann, das Lernen schwerer zu machen—Kognitive Aktivierung und die Kraft des Vergleichens. In K. Schöppe & F. Schulz (Hrsg.), Kreativität & Bildung—Nachhaltiges Lernen, 77–132. kopaed.

Lipowsky, D., Drollinger-Vetter, B., Klieme, E., Pauli, C. & Reusser, K. (2018). Generische und fachdidaktische Dimensionen von Unterrichtsqualität –Zwei Seiten einer Medaille? In M. Martens, K. Rabenstein, K. Bräu, M. Fetzer, H. Gresch, I. Hardy & C. Schelle (Hrsg.), Konstruktionen von Fachlichkeit: Ansätze, Erträge und Diskussionen in der empirischen Unterrichtsforschung, 183–202. Klinkhardt.

Lotz, M., Gabriel, K., & Lipowsky, F. (2013). Niedrig und hoch inferente Verfahren der Unterrichtsbeobachtung. Analysen zu deren gegenseitiger Validierung. Zeitschrift für Pädagogik, 59(3), 357–380.

Mainhard, M.Tim/Pennings, Helena J.M./Wubbels, Theo/Brekelmans, Mieke (2011): Mapping control and affiliation in teacher student interaction with State Space Grids. In: Teaching and Teacher Education Volume 28, Issue 7 2012, 1027–1037.

Lunetta, V.N. (1998): The school science laboratory: Historical perspectives and contexts for contemporary teaching. In: Tobin, K./Fraser, B. (Eds.): International Handbook of Science Education, 249–262 Kluwer.

Maskus, R. (1976). Unterricht als Prozess. Dynamisch-integratives Strukturmodell. Bad Heilbrunn.

Maor, E. (2019). The Pythagorean theorem: a 4,000-year history. Princeton University Press.

Mayr, J. (2006). Klassenführung auf der Sekundarstufe II: Strategien und Muster erfolgreichen Lehrerhandelns. Swiss Journal of Educational Research, 28(2), 227–242.

Merzyn, G. (2015). Guter Physikunterricht. Die Sicht von Schülern, Lehrern und Wissenschaftlern. PhyDid B-Didaktik der Physik-Beiträge zur DPG-Frühjahrstagung.

Montgomery, H., & Adelbratt, T. (1982). Gambling decisions and information about expected value. Organizational Behavior and Human Performance, 29(1), 39–57.

Mörters, P., & Peres, Y. (2010). Brownian motion (Vol. 30). Cambridge University Press.

Neuber, V., Gebhard, C. & Lipowsky, F. (2015, September). Unterschiede in der Unterrichtsqualität aus Schülersicht: Eine Frage des Lehrer-Typs. Poster auf der 80. Tagung der Arbeitsgruppe für Empirische Pädagogische Forschung (AEPF).

Neuenschwander, M. P. (2006). Klassenführung –Konzepte und neue Forschungsbefunde. Schweizerische Zeitschrift für Bildungswissenschaften, 28(2), 189–203.

Oltmanns, T. F., Friedman, J. N., Fiedler, E. R., & Turkheimer, E. (2004). Perceptions of people with personality disorders based on thin slices of behavior. Journal of Research in Personality, 38, 216–229.

Papula, L. (2017). Mathematische Formelsammlung (Vol. 4). Springer Fachmedien Wiesbaden.

Pellicer-Sánchez, A., Tragant, E., Conklin, K., Rodgers, M., Serrano, R., & Llanes, Á. (2020). YOUNG LEARNERS'PROCESSING OF MULTIMODAL INPUT AND ITS IMPACT ON READING COMPREHENSION: AN EYE-TRACKING STUDY. Studies in Second Language Acquisition, 42(3), 577–598.

Pennings, Helena J.M., Hollenstein, Tom (2019): Teacher-Student Interactions and Teacher Interpersonal Styles: A State Space Grid Analysis. In: The Journal of Experimental Education vom 07.04.2019, o. S, https://www.tandfonline.com/doi/full/10.1080/00220973.2019.1578724.

Pennings, Helena J. M., Mainhard, Tim (2016): Analyzing Teacher-Student Interactions with State Space Grids. In: Complex Dynamical Systems in Education, 233–271.

Pennings, Helena J.M./van Tartwijk, Jan/Wubbels, Theo/Claessens, Luce C.A./van der Want, Anna C./ Brekelmans, Mieke (2014): Real-time teacher student interactions: A Dynamic Systems approach. In: Teaching and Teacher Education Volume 37, 2014, 183–193.

Pennings, H. J., Brekelmans, M., Wubbels, T., van der Want, A. C., Claessens, L. C., & van Tartwijk, J. (2014). A nonlinear dynamical systems approach to real-time teacher behavior: Differences between teachers. Nonlinear Dynamics, Psychology, and Life Sciences, 18(1), 23–45.

Peng, J., Zhang, H., & Wang, D. (2018). Measurement and analysis of teaching and background noise level in classrooms of Chinese elementary schools. Applied Acoustics, 131, 1–4.

Praetorius, A./Klieme, E./Herbert, B./Pinger, P. (2018): Generic dimensions of teaching quality: the German framework of Three Basic Dimensions. In: ZDM Mathematics Education June 2018, Volume 50, Issue 3, 407–426.

Praetorius, A., Rogh, W., Kleickmann, T. (2020). Blinder Fleck des Modells der drei Basisdimensionen von Unterrichtsqualität? Das Modell im Spiegel einer internationalen Synthese von Merkmalen der Unterrichtsqualität. Wiesbaden: Springer Fachmedien.

Praetorius, A. K., Herrmann, C., Gerlach, E., Zülsdorf-Kersting, M., Heinitz, B., & Nehring, A. (2020). Unterrichtsqualität in den Fachdidaktiken im deutschsprachigen Raum-zwischen Generik und Fachspezifik. Unterrichtswissenschaft, 48(3), 409–446.

Pietsch, M. (2010). Evaluation von Unterrichtsstandards. Zeitschrift für Erziehungswissenschaft, 13, 121–148.

Rakoczy, K. (2008). Motivationsunterstützung im Mathematikunterricht. Unterricht aus der Perspektive von Lernenden und Beobachtern. Waxmann.

Rule, N. O., & Ambady, N. (2008). Brief exposures: Male sexual orientation is accurately perceived at 50ms. Journal of Experimental Social Psychology, 44, 1100–1105.

Ryan, R. M. & Deci, E. L. (2000). Self-determination theory and the facilitation of intrinsic motivation, social development, and well being. American Psychologist, 55(1), 68–78.

Samek, A. (2019): Advantages and disadvantages of field experiments. In Handbook of Research Methods and Applications in Experimental Economics. Cheltenham, UK: Edward Elgar Publishing

Sandelius, M. (1967). The Teacher's Corner: A Note on the Variance of a Discrete Uniform Distribution. The American Statistician, 21(5), 21–21.

Schlesinger, L., Jentsch, A., Kaiser, G., König, J., & Blömeke, S. (2018). Subject-specific haracteristics of instructional quality in mathematics education. ZDM, 50(3), 475–490.

Scherer, K. R. (1972). Judging personality from voice: A cross-cultural approach to an old issue in interpersonal perception. Journal of Personality, 40, 191–210.

Scherzinger, Marion, Roth, Benjamin & Wettstein, Alexander (2020). Pädagogische Interaktionen als Grundbaustein der Lehrperson-Schüler*innen-Beziehung. Die Erfassung mit State Space Grids. Unterrichtswissenschaft, 49(3), 303–324.

Schindler-Tschirner, S., & Schindler, W. (2019). Summieren leicht gemacht. In Mathematische Geschichten II-Rekursion, Teilbarkeit und Beweise, 9–12. Springer Spektrum.

Schirotzek, Winfried/Scholz, Siegfried (2005): Starthilfe Mathematik. Für Studienanfänger der Ingenieur-, Natur- und Wirtschaftswissenschaften. Springer-Verlag.

Seidel, T. (2009): Pädagogische Psychologie. Hrg. von Wild, E. u. Möller, J., Springer Medizin.

Seidel, T. & Shavelson, R. J. (2007). Teaching effectiveness research in the past decade: Role of theory and research design in disentangling meta-analysis results. Review of Educational Research, 77(4), 454–499.

Syring, M. (2017): Classroom Management Theorien, Befunde, Fälle—Hilfen für die Praxis. Vandenhoeck & Ruprecht.

Tamar, A., Di Castro, D., & Mannor, S. (2016). Learning the variance of the reward-to-go. The Journal of Machine Learning Research, 17(1), 361–396.

Van Dongen, Peter (2015): Einführungskurs Mathematik und Rechenmethoden. Springer.

Udo Kuckartz & Stefan Rädiker (2022): Qualitative Inhaltsanalyse. Methoden, Praxis, Computerunterstützung. Weinheim Beltz Juventa Grünwald Preselect.media GmbH 2022.

van de Craats, J., & Bosch, R. (2010). Parametrisierte Kurven. In Grundwissen Mathematik, 166–173. Springer

Viennot, L. (2021). Incomplete explanations in physics teaching: discussing the rainbow with student teachers. European Journal of Physics, 42(5), 055705.

Wang, M. Z., Chen, K., & Hall, J. A. (2021). Predictive Validity of Thin Slices of Verbal and Nonverbal Behaviors: Comparison of Slice Lengths and Rating Methodologies. Journal of Nonverbal Behavior, 45(1), 53–66.

Wannack, E. (2012). Classroom Management und seine Bedeutung für die Gestaltung von Spiel- und Lernaktivitäten. In F. Hellmich, S. Förster & F. Hoya (Hrsg.), Bedingungen des Lehrens und Lernens in der Grundschule. Bilanz und Perspektiven, 69–72. Springer.

Wirth, J., Thillmann, H., Künsting, J., Fischer, H. E., & Leutner, D. (2008). Das Schülerexperiment im naturwissenschaftlichen Unterricht. Bedingungen der Lernförderlichkeit einer verbreiteten Lehrmethode aus instruktionspsychologischer Sicht. Zeitschrift für Pädagogik, 54(3), 361–375.

Wubbels, T., Brekelmans, M., den Brok, P., Wijsman, L., Mainhard, T., & van Tartwijk, J. (2015). Teacherstudent relationships and classroom management. In E. T. Emmer & E. J. Sabornie (Hrsg.), Handbook of classroom management, 363–386. Routledge.

Yildirim, I. (2015). A study on physical education teachers: The correlation between self-efficacy and job satisfaction. Education, 135(4), 477–485.

Zhao, F., Schnotz, W., Wagner, I., & Gaschler, R. (2014). Eye Tracking Indicators of Reading Approaches in Text-Picture Comprehension. Frontline Learning Research, 2(5), 46–66.

Zuckerman, J. T. (2007). Classroom management in secondary schools: A study of student teachers' successful strategies. American secondary education, 4–16.

Printed in the United States
by Baker & Taylor Publisher Services